WITHDRAWN
UTSA Libraries

RENEWALS 458-4574
DATE DUE

RESEARCH AND PERSPECTIVES IN ALZHEIMER'S DISEASE

Dennis J. Selkoe Antoine Triller Yves Christen (Eds.)

Synaptic Plasticity and the Mechanism of Alzheimer's Disease

With 34 Figures, 28 in color

Dr. Dennis J. Selkoe
Center for Neurologic Diseases
Brigham and Women's Hospital
Harvard Medical School
Boston, MA 02115, USA
email: dselkoe@rics.bwh.harvard.edu

Dr. Antoine Triller
Laboratoire de Biologie Cellulaire de la Synapse
Institut National de la Santé et de la Recherche Médicale Unite 789
Ecole Normale Supérieure, 46 rue d'Ulm
75005 Paris
France
email: triller@biologie.ens.fr

Dr. Yves Christen
Fondation IPSEN
Pour la Recherche Thérapeutique
24, rue Erlanger
75781 Paris Cedex 16
France
email: yves.christen@ipsen.com

ISBN 978-3-540-76329-1 e-ISBN 978-3-540-76330-7

DOI 10.1007/978-3-540-76330-7

Library of Congress Control Number: 2007938204

© 2008 Springer-Verlag Berlin Heidelberg

This work is subject to copyright. All rights are reserved, whether the whole or part of the material is concerned, specifically the rights of translation, reprinting, reuse of illustrations, recitation, broadcasting, reproduction on microfilm or in any other way, and storage in data banks. Duplication of this publication or parts thereof is permitted only under the provisions of the German Copyright Law of September 9, 1965, in its current version, and permission for use must always be obtained from Springer. Violations are liable to prosecution under the German Copyright Law.

The use of general descriptive names, registered names, trademarks, etc. in this publication does not imply, even in the absence of a specific statement, that such names are exempt from the relevant protective laws and regulations and therefore free for general use.

Typesetting: LE-TEX Jelonek, Schmidt & Vöckler GbR, Leipzig, Germany
Production: LE-TEX Jelonek, Schmidt & Vöckler GbR, Leipzig, Germany
Cover design: WMX Design GmbH, Heidelberg

Printed on acid-free paper

9 8 7 6 5 4 3 2 1

springer.com

Library
University of Texas
at San Antonio

Preface

Remarkable progress has been made in Alzheimer's research over the last decade. Our understanding of which molecular mechanisms are involved in the cascade of events characteristic of the disease is growing. However, the mechanisms by which they induce deleterious effect are still not clear. Many recent studies have confirmed that the synapse is a key target in the pathogenesis of Alzheimer's disease. Not only does synaptic loss occur at an early stage of the disease, but it seems that β-amyloid peptide deposits or microtubular disorders affect synaptic function. This may explain the memory deficits that are so characteristic of the disease. It is therefore logical to investigate therapeutic approaches targeting the deleterious mechanisms that affect the synapse. This area of research promises to be all the more productive because synaptic dysfunction is involved in various other neurodegenerative diseases or, more generally speaking, central nervous system disorders.

This new awareness has emerged at a time when neuroscientific research is uncovering more about synaptic function and the multiple proteins it involves; hence our decision to invite both fundamental researchers in synaptic biology and specialists in neurogenerative disease to the Colloque Médecine et Recherche organized by the Fondation Ipsen in Paris on April 16, 2007.

D.J. Selkoe, A. Triller and Y. Christen

Acknowledgements. The editors want to thank Jacqueline Mervaillie, Astrid de Gérard and Sonia Le Cornec for the organization of this meeting and Mary Lynn Gage for the editing of the book.

Table of Contents

List of Contributors

Alvarez, Veronica A.
Department of Neurobiology, Harvard Medical School, Boston MA 02115, USA

Anwyl, Roger
Department of Physiology, and Trinity College Institute of Neuroscience, Trinity College Dublin 2, Ireland

Bloodgood, Brenda L.
Department of Neurobiology, Harvard Medical School, Boston MA 02115, USA

Bloom, Floyd E.
Molecular and Integrative Neuroscience Department, The Scripps Research Institute, La Jolla CA 92037, USA

Bourgeron, Thomas
Institut Pasteur, Université Denis Diderot Paris 7, 25 rue du Docteur Roux, 75724 Paris Cedex 15, France

Bredesen, Dale E.
Buck Institute for Age Research, Novato, CA 94945 and Department of Neurology, University of California, San Francisco CA 94143, USA

Cirrito, John R.
Department of Psychiatry, Neurology, Molecular Biology and Pharmacology, Hope Center for Neurological Disorders, Washington University School of Medicine, St. Louis, MO 63110, USA

Cullen, William K.
Department of Pharmacology & Therapeutics, Biotechnology Building, Trinity College Dublin 2, Ireland

De Felice, Fernanda
Department of Neurobiology and Physiology, Cognitive Neurology and Alzheimer's Disease Center, Northwestern University, Evanston, IL 60208

Esteban, José A.
Department of Pharmacology, University of Michigan Medical School, Ann Arbor, MI 48109-0632, USA

De Paola, V.
Cold Spring Harbor laboratory, Cold Spring Harbor, New York 11724, USA
MRC-Clinical Sciences Centre, Imperial College London, London W12 0NN, United Kingdom

Holtmaat, Antony
Cold Spring Harbor laboratory, Cold Spring Harbor, New York 11724, USA
Départment des Neurosciences fondamentales, University of Geneva, 1211 Geneve, Switzerland

Holtzman, David M.
Department of Neurology, Molecular Biology and Pharmacology, Hope Center for Neurological Disorders, Washington University School of Medicine, St. Louis, MO 63110, USA

Hsieh, Helen
Watson School of Biological Sciences, Cold Spring Harbor Laboratory, New York, NY 11724-2213, USA

Hu, NengWei
Department of Pharmacology & Therapeutics, Biotechnology Building, Trinity College Dublin 2, Ireland

Klein, William L.
Department of Neurobiology and Physiology, Cognitive Neurology and Alzheimer's Disease Center, Northwestern University, Evanston, IL 60208

Klyubin, Igor
Department of Pharmacology & Therapeutics, Biotechnology Building, Trinity College Dublin 2, Ireland

Knott, Graham
Départment de Biologie Cellulaire et de Morphologie, University of Lausanne, 1005 Lausanne, Switzerland

Koo, Edward H.
Department of Neurosciences, University of California, San Diego, La Jolla CA 92037, USA

Lacor, Pascale N.
Department of Neurobiology and Physiology, Cognitive Neurology and Alzheimer's Disease Center, Northwestern University, Evanston IL 60208, USA

Lambert, Mary P.
Department of Neurobiology and Physiology, Cognitive Neurology and Alzheimer's Disease Center, Northwestern University, Evanston IL 60208, USA

Malinow, Roberto
Cold Spring Harbor Laboratory, Cold Spring Harbor, New York NY 11724, USA
State University of New York and Stonybrook, Dept. Neurobiology, USA
Watson School of Biological Sciences, Cold Spring Harbor Laboratory, USA

Mennerick, Steven
Department of Psychiatry, Anatomy, and Neurolobiology, Hope Center for Neurological Disorders, Washington University School of Medicine, St. Louis, Missouri, 63110, USA

Renner, Marianne
Laboratoire de Biologie Cellulaire de la Synapse, Institut National de la Santé et de la Recherche Médicale Unite 789, Ecole Normale Supérieure, 46 rue d'Ulm, 75005 Paris, France

Rowan, Michael J.
Department of Pharmacology & Therapeutics, Biotechnology Building, Trinity College Dublin 2, Ireland

Sabatini, Bernardo L.
Department of Neurobiology, Harvard Medical School, Boston MA 02115, USA

Selkoe, Dennis J.
Center for Neurologic Diseases, Brigham and Women's Hospital and Harvard Medical School, Boston MA 02115, USA

Shankar, Ganesh M.
Department of Neurobiology, Harvard Medical School, Boston MA 02115, USA
Center for Neurologic Diseases, Brigham and Women's Hospital and Harvard Medical School, Boston MA 02115, USA

Sheng, Morgan
The Picower Institute for Learning and Memory, Howard Hughes Medical Institute, Depts of Brain and Cognitive Sciences, and Biology, Massachusetts Institute of Technology, 77 Massachusetts Avenue (46-4303) Cambridge MA 02139, USA

Stewart, Floy R.
Department of Neurology, Hope Center for Neurological Disorders, Washington University School of Medicine, St. Louis, MO 63110, USA

Triller, Antoine
Laboratoire de Biologie Cellulaire de la Synapse, Institut National de la Santé et de la Recherche Médicale Unite 789, Ecole Normale Supérieure, 46 rue d'Ulm, 75005 Paris, France

Wei, Wei
State University of New York and Stony Brook, Department of Neurobiology, New York NY 11794-80-34, USA

Wilbrecht, L.
Cold Spring Harbor laboratory, Cold Spring Harbor, New York NY 11724, USA
Keck Center, UCSF, San Francisco CA 94143, USA

Zhao, Wei-Qin
Department of Neurobiology and Physiology, Cognitive Neurology and Alzheimer's Disease Center, Northwestern University, Evanston IL 60208, USA

Permanence of the Synapse and Molecular Instability

Marianne Renner[1] and *Antoine Triller*[1]

Summary. A complex molecular assembly accounts for receptor accumulation at synapses and "synaptic plasticity" derives partly from modifications of postsynaptic receptor number resulting from receptor trafficking. New concepts now emerge from imaging of receptor movements at the single molecule level. Inhibitory glycine or GABA receptors, and the excitatory AMPA and NMDA glutamate receptors are mobile in synapses, switch between extrasynaptic and synaptic localizations by lateral diffusion and can be exchanged between synapses through lateral diffusion in the plane of the extrasynaptic plasma membrane. This dynamic behavior can be tuned by the cytoskeleton or by the synaptic activity. Variations in receptor numbers at postsynaptic sites is therefore likely to depend, not only on regulated exo-endocytotic processes but also on regulation of diffusion by modification of the structure of the membrane and/or by transient interactions with scaffolding proteins. This diffusive behavior provides a new framework for the understanding of synaptic strength regulations. Because receptors diffuse in the plasma membrane, in the same way that particles do in a two-dimensional field, they could be transiently trapped at specific loci corresponding to postsynaptic densities. We will present new data on the regulation of the diffusive properties of inhibitory receptors by excitatory activity. This open new routes towards understanding not only the dynamic equilibrium accounting for receptor number at synapses, but also the mechanisms underlying the excitation-inhibition balance during modifications induced by so-called plasticity within neuronal networks.

Chemical synapses are specialized contacts between neurons across which information is transmitted by directed secretion. Both the pre- and postsynaptic sides have precise molecular compositions and organizations. On the postsynaptic side, responses to the release of neurotransmitters are generated by a set of receptors and related proteins embedded in the so-called postsynaptic density (PSD). The latter has been characterized by electron microscopy as an electron-dense fuzzy meshwork below the plasma membrane and facing the presynaptic active zone. PSDs comprise several hundred different proteins with a defined spatial organization (reviewed in Sheng and Hoogenraad 2006). Fast excitation and inhibition rely upon the presence of ionotropic receptors present at high density in front of the presynaptic terminal. These receptors are concentrated at the postsynaptic membrane through direct and indirect interactions with specific scaffolding proteins and cytoskeletal elements (Fig. 1).

The modifications of synaptic strength as a consequence of previous activity are known as synaptic plasticity, which is thought to be the basis for the adaptation of the nervous system to a constantly changing environment and for keeping track of previous experiences. In particular, modification of the number of postsynaptic receptors is

[1] Laboratoire de Biologie Cellulaire de la Synapse, Institut National de la Santé et de la Recherche Médicale Unite 789, Ecole Normale Supérieure, 46 rue d'Ulm, 75005 Paris, France,
email: triller@biologie.ens.fr

Selkoe et al.
Synaptic Plasticity and the Mechanism of Alzheimer's Disease
© Springer-Verlag Berlin Heidelberg 2008

one of the mechanisms responsible for the long-term potentiation (LTP) or long-term depression (LTD) that may underlie learning and memory (reviewed in Sheng and Kim 2002).

Two apparently contradictory facts have to be reconciled: one is the continuous exchange of synaptic components in and out of the postsynaptic membrane (Choquet and Triller 2003) and the other is the amazing persistence of synaptic structures, which can last months – maybe years – in the central nervous system (Holtmaat et al. 2006). In this review, we will 1) briefly describe the molecular organization of ionotropic receptors and their stabilization at synapses, 2) describe the turnover and diffusion dynamics of excitatory and inhibitory receptors, and 3) propose a framework to reconcile synaptic stability and molecular dynamics.

The Neuronal Synapse: Structure and Molecular Components

Inhibitory Contacts

Fast synaptic inhibition is mediated by two types of chloride-permeable ion channels: the γ-aminobutyric acid receptor (GABAR) and the glycine receptor (GlyR). Inhibitory synapses can be GABAergic, glycinergic or mixed (i.e., respond to both neurotransmitters). The identity of the presynaptic element is the determining factor for the composition of the postsynaptic membrane that is enriched in one type of receptor or both (Levi et al. 1999; Dugué et al. 2005).

GlyR was the first receptor in the central nervous system that was shown to form microdomains (Triller et al. 1985), which are stabilized at the PSD by the scaffolding protein gephyrin (Kirsch et al. 1993). GlyRs contain five subunits: two α and three β (reviewed in Legendre 2001; Vannier and Triller 2007). The β subunit mediates the interaction with gephyrin (Meyer et al. 1995). Similarly, the ionotropic GABA receptors are stabilized at synapses. They can be differentiated into $GABA_AR$ and $GABA_CR$ on the basis of their pharmacology. $GABA_CR$ is almost exclusively expressed in the adult retina. It is composed of the ϱ-type subunit, which has three isoforms. $GABA_AR$ pentamers display high structural heterogeneity, with 21 subunit types belonging to eight different families. It is thought that most functional $GABA_ARs$ in vivo are formed by α, β, and γ subunits (reviewed in Fritschy et al. 2003).

Gephyrin is the core protein of the scaffolding multimolecular complex in inhibitory contacts. Antisense experiments and studies with knockout mice have revealed that gephyrin is required for the synaptic localization of GlyR (Kirsch et al. 1993; Feng et al. 1998). Briefly, gephyrin is composed of three domains named G, E and L. The N-terminal G and C-terminal E domains are homologous to two bacterial proteins involved in the biosynthesis of the molybdenum cofactor (Feng et al. 1998; Ramming et al. 2000). The binding site of the GlyR β subunit has been mapped on a gephyrin-specific structure in the E region crystal (Sola et al. 2004; Kim et al. 2006). Isolated G and E domains form stable trimers and dimers. With these binding properties gephyrin may build up a two-dimensional hexagonal lattice (Kneussel and Betz 2000). Actually, gephyrin trimerization correlates well with its ability to stabilize GlyR at the cell surface. In cultured neurons, the incorporation of a trimerization-defective gephyrin molecule into the postsynaptic gephyrin polymer interferes with the presence of GlyR at synapses (Bedet et al. 2006).

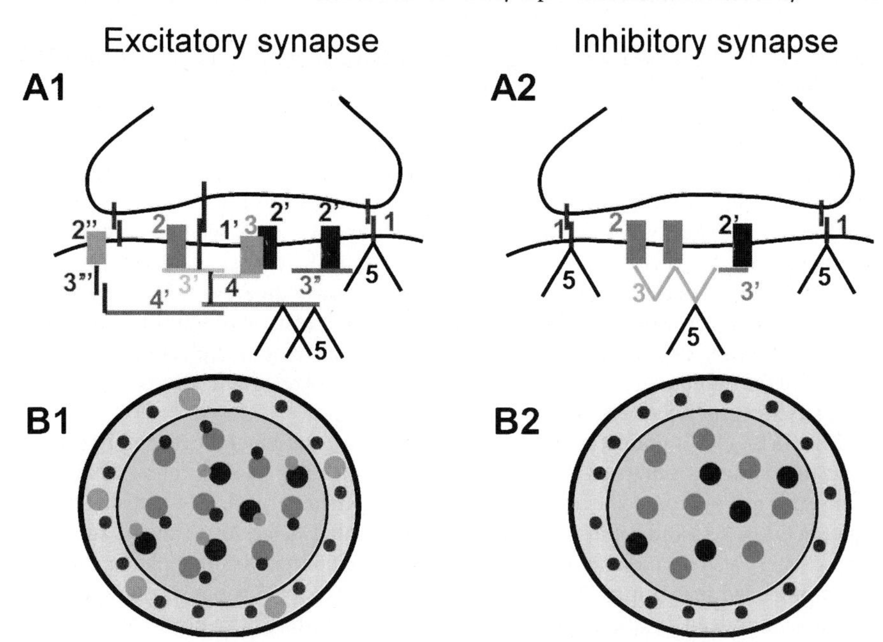

Fig. 1. Schematic representation of the structural organization of generic glutamatergic excitatory (left) and inhibitory (right) synapses. **A.** Transverse view of synaptic contacts with emphasis on the postsynaptic scaffold network organization. **A1.** Organization of PSD at excitatory synapses. Transmembrane proteins include adhesive proteins such as cadherin-catenin complexes (1) at the periphery of the synapse, neuroligin (1′), as well as NMDA (2), AMPA (2′) or metabotropic (2″) glutamate receptors, and the AMPA receptor interacting protein stargazin (3). In a first layer, the postsynaptic scaffold includes PSD-95 (3′), which interacts with NMDAR, neuroligin and stargazin, GRIP (3″), which interacts with AMPAR, and Homer (3‴), which binds to mGluR. In a second layer, PSD-95 binds to GKAP (4), which in turn connects to Shank (4′). The cytoskeleton (5) interacts at various levels with the postsynaptic synaptic molecular network. **A2.** Organization of PSD at inhibitory synapses. Transmembrane proteins include adhesive proteins such as cadherins (1) at the periphery of the synapse. Glycine (2) and GABA (2′) receptors bind directly and indirectly, respectively, through an unidentified intermediate molecule (3′) to gephyrin (3). **B.** Top view of the synapse. The cadherins are concentrated on an annulus around both excitatory and inhibitory synapses. Neuroligins are only detected at excitatory synapses. Transmembrane proteins are shown with the same color code as in A (reprinted with permission from Triller and Choquet, 2003)

The L domain is highly variable and several gephyrin splice variants have been identified (Ramming et al. 2000), but the affinity of gephyrin for GlyR is not directly related to L variability (Bedet et al. 2006). However, the variation in its primary structure is correlated with numerous gephyrin-interacting proteins (Kneussel and Betz 2000). Indeed, gephyrin binds directly to microtubules and this interaction is critical for the packing of receptors at synaptic contacts (Kirsch and Betz 1995). Other recently reported binding partners suggest that gephyrin could be the core of a platform and it may also be implicated in modulating signaling pathways and the actin cytoskeleton. These partners comprise the GDP/GTP exchange factor collybistin and several cytoskeleton regulators such as profilin, debrin and mammalian enabled (Mena)/vasodilator-stimulated phosphoprotein (VASP; Mammoto et al. 1998; refs. in Moss and Smart 2001; Giesemann et al. 2003). Raft1 also binds to gephyrin (Sabatini et al. 1999) and interestingly is implicated in the control of mRNA translation. It could therefore be involved in the synaptic regulation of GlyR synthesis from mRNAs present in dendrites (Racca et al. 1997). Collibistin is a GTP/GDP exchange factor (GEF) that enhances the Cdc42 rac/rho-like GTPase. It may participate in controlling the subsynaptic cytoskeleton, since Cdc42 is implicated in the regulation of actin polymerization (Kins et al. 2000; refs. in Moss and Smart 2001; Lüscher and Keller 2004). Dynein light chains 1 and 2 (Dlc 1 and 2) were also identified as gephyrin-binding partners by yeast two-hybrid experiments (Fuhrmann et al. 2002). This association suggests that dynein could be involved in regulating the number of gephyrin molecules at synapses (reviewed in Moss and Smart 2001; Lüscher and Keller 2004).

The $GABA_A R$ $\gamma 2$ subunit plays a role in synaptic localization of the receptor because its accumulation at synapses is impaired in knockout mice (Essrich et al. 1998). Although this role appears to be mediated by gephyrin (Kneussel et al. 1999), no direct link between $GABA_A Rs$ and gephyrin has been demonstrated so far. As $\alpha 2$ and $\gamma 2$ subunits form clusters at pyramidal synapses in hippocampal cultures from gephyrin-/-mice, it was proposed that some $GABA_A Rs$ cluster at synapses by a gephyrin-independent mechanism. Indeed, in immature synapses, $GABA_A R$ clustered at a postsynaptic location prior to gephyrin and GlyR (Dumoulin et al. 2000). $GABA_A R$ could thus be the initiating element in the construction of mixed glycine and GABAergic synapses. Therefore, it was proposed that gephyrin could be involved in the stabilization of $GABA_A R$ rather than in its initial accumulation at synaptic sites. Once clustered, $GABA_A R$ would recruit gephyrin, which might then nucleate the other components of the inhibitory PSD (Levi et al. 2004). The synaptic targeting of $GABA_A Rs$ in the absence of gephyrin could be mediated by the palmitoylation of the $\gamma 2$ subunit (Rathenberg et al. 2004).

Various proteins that directly interact with GABA receptors have been identified by yeast two-hybrid and co-immunoprecipitation experiments. They include GABARAP (GABA receptor-associated protein), Raft1 and Plic-1 (reviewed in Moss and Smart 2001). GABARAP binds to the $\gamma 2$ subunit and to gephyrin as well as to the N-ethylmaleimide-sensitive factor (NSF). Many lines of evidence indicate that it is probably involved in the intracellular trafficking of the receptor rather than in its stabilization at synapses (Kittler et al. 2001; reviewed in Moss and Smart 2001). Plic-1 is a ubiquitin-like protein that is enriched at inhibitory synapses and associates with subsynaptic membranes (Bedford et al. 2001). The function of Plic-1 is to increase the number of receptors available for insertion in the plasma membrane by increasing the stability of intracellular GABAR as a consequence of the inhibition of its polyubiquitination (Bedford et al. 2001). Another protein that is co-clustered with $GABA_A$ receptors is dystrophin (Levi et al. 2002), but a direct interaction between them has not been proven. Interestingly, the absence of dystrophin leads to a reduction in $GABA_A Rs$ but not in gephyrin clusters (Knuesel et al. 1999).

Excitatory Contacts

Synaptic excitation is mostly mediated by glutamate, which activates ionotropic (AMPA type or NMDA type) or G-protein coupled receptors (mGluR). The NMDA receptor

(NMDAR) is composed of two obligatory NR1 subunits (with eight splice variants) and up to two NR2 subunits (NR2A–NR2D). NMDARs are associated with different scaffolding molecules of the membrane-associated guanylate kinase (MAGUK) family. More specifically, NR2A-rich receptors are linked to postsynaptic density-95 (PSD-95), whereas NR2B-rich receptors are coupled to synapse associated protein-102 (SAP-102; ref. in van Zundert et al. 2004). AMPARs are mainly tetramers of GluR1/GluR2 but can also be formed by GluR3/GluR2 subunits (refs. in Derkach et al. 2007). Although these subunits are highly homologous, they differ at the carboxyl (C) termini, which contain regulatory domains that are targeted by multiple signal transduction pathways. The C termini also interact with different scaffolding proteins, so AMPARs are stabilized by specific proteins depending on their subunit composition (reviewed in Derkach et al. 2007).

Scaffolding elements are mainly proteins belonging to the PDZ family that interact with receptors via their PDZ [postsynaptic density-95 (PSD-95)/discs large (Dlg)/zona occludens-1 (ZO-1)] domain. PDZ-containing proteins include, for example, MAGUK family members, GRIP and PICK (reviewed in Kim and Sheng 2004). In addition, TARPs (transmembrane AMPAR regulatory proteins) play a fundamental role in modulating the activity of AMPA receptors and linking them to the postsynaptic scaffold (Nicoll et al. 2006). The TARP stargazin interacts directly with PSD-95 and AMPAR and controls the number of receptors at synapses (reviewed in Nicoll et al. 2006). These "first-line" proteins are coupled to other members of the scaffold assembly. For example, PSD-95 binds to GKAP (guanylate kinase-associated protein), which interacts with Shank, which in turn binds to Homer. Scaffolding molecules also assemble signaling cascades and enzyme complexes that have fundamental functions in synaptic transmission (reviewed in Kim and Sheng 2004).

In addition to classic biochemical analysis, mass spectrometry approaches have been used to identify and quantify proteins in purified PSD fractions or in immunoprecipitated NMDAR complexes. Different sets of proteins with many different functions are present at the excitatory PSD. They include the most obvious ones, such as receptors and scaffolding components, but also members of signaling pathways, regulators of ubiquitination, RNA trafficking, protein translation and others (Jordan et al. 2004; Peng et al. 2004; reviewed in Sheng and Hoogenraad 2007).

Using quantitative mass spectrometry (absolute quantitation mass spectrometry or AQUA MS), the molar concentration and relative stoichiometries of some key components of the PSDs were measured (Peng et al. 2004). As expected, PSD-95 is highly abundant in the adult rat forebrain (~1% by mass), with ~300 copies on average (Chen et al. 2005). Other proteins quantified were GKAP (~150 copies), Shank (~150 copies) and Homer (~60 copies; Cheng et al. 2006). PSD-95 probably participates in other scaffolding families, as it can interact with many other proteins via its PDZ domains (Kim and Sheng 2004). Calmodulin kinase II (CaMKII) isoforms α and β are the most abundant proteins in the PSD representing ~7.4% and 1.3% in mass, respectively. The abundance of CaMKIIs, which were shown to form an intracellular network (Petersen et al. 2003), points to a possible scaffolding role in addition to their kinase activity.

Interestingly, receptors are present in much lesser amounts at the PSD. The number of NMDA receptors, which are considered to be the functional core of the PSD, was found to be ~20 (Peng et al. 2004). Shigemoto and colleagues applied a highly sensitive, freeze-fracture replica labeling technique and electrophysiological studies to quantify the amount of functional AMPARs. In immature rat Purkinje cells, the density of synaptic receptors is ~1 280 receptors/μm^2 and the number of receptors at a given synapse is proportional to the area of the PSD (Tanaka et al. 2005). The high sensitivity of this technique allowed the measurement of AMPAR density at the extrasynaptic membrane, which was found to be 50-fold lower than at the PSD. Further studies of other excitatory synapses of the cerebellum revealed a high degree of heterogeneity between different synapses in terms of AMPAR density and distribution, which may reflect distinct regulation of individual synaptic efficacy through receptor turnover (Masugi-Tokita et al. 2007).

Within the PSD, the distribution of AMPARs and NMDARs can be uniform or in small clusters and probably depends upon the type of the presynaptic afferent button. AMPA receptors are either distributed uniformly (Nusser et al. 1994) or accumulate at the edge of synaptic complexes (Matsubara et al. 1996). In the cerebellum, synapses established between parallel fibers and Purkinje cells displayed clusters of AMPARs whereas climbing fibers – Purkinje cell synapses and parallel fibers – interneuron synapses did not (Masugi- Tokita et al. 2007). The NMDA receptors are either evenly distributed at the PSDs (Clarke and Bolam 1998) or concentrated at its center (Somogyi et al. 1998). In contrast, metabotropic glutamate receptors (mGluRs) are concentrated in an annulus surrounding the postsynaptic differentiation (Matsubara et al. 1996; Baude et al. 1993; Lujan et al. 1997). However, the amount of mGluR is still high at a distance from the PSD and 75% of the total surface immunoreactivity is on non-synaptic areas (Lujan et al. 1997). These different patterns suggest the existence of a synaptic-specific microorganization.

Turnover and Trafficking of Synaptic Components

Since the initial ultrastructural localization of gephyrin at the central synapse (Triller et al. 1985), numerous scaffolding molecules have been described in the CNS, reinforcing the concept that receptors are concentrated due to their stabilization at the postsynaptic membrane. In other words, the accepted notion was that receptors were fixed at synapses and that this accounted for their accumulation. However, evidence of rapid trafficking and recycling of receptors was found for both the inhibitory and excitatory synapses. More importantly, the dynamics of receptors are regulated by synaptic activity (reviewed in Kittler and Moss 2003; Derkach et al. 2007). Actually, the number of receptors at the membrane surface is determined by the ratio between exocytosis and endocytosis. Internalized receptors can be recycled to the surface or be targeted to degradation (reviewed in Triller and Choquet 2005).

The stability of $GABA_A$ receptors at the cell surface seems to be regulated by phosphorylation. $GABA_A$R subunits are phosphorylated by cAMP-dependent protein kinases A and C (PKA and PKC), cGMP-dependent protein kinase (PKG), CaMKII and the tyrosine kinase Src (reviewed in Moss and Smart 2001). There is evidence that the number of $GABA_A$Rs at the plasma membrane can be increased by insulin and reduced by brain-derived neurotrophic factor (BDNF; reviewed in Kittler and Moss 2003; Lüscher and Keller 2004).

Much less is known about trafficking and turnover of the glycine receptor. Long-term inhibition by strychnine provokes its disappearance from synapses and sequestration in a non-endocytic compartment (Levi et al. 1998; Rasmussen et al. 2002). Interestingly, GlyR associates with gephyrin intracellularly and this association increases the accumulation of receptors at the plasma membrane (Hanus et al. 2004). Using cleavable tags and transfection experiments, it was shown that a large proportion of newly synthesized GlyRs are inserted at non-synaptic sites in the plasma membrane of the cell body and the initial portion of dendrites. They subsequently travel along neurites by lateral diffusion at a linear speed of about 1 μm/min (Rosenberg et al. 2001).

AMPARs are constantly recycling, and this process is controlled by neuronal activity and several proteins like GRIP, PICK and NSF, which interact with the C-terminus of distinctive AMPAR subunits and regulate different steps of trafficking (reviewed in Derkach et al. 2007). In particular, receptors containing the GluR2 subunit spend between 10 and 30 minutes at the plasma membrane (refs. in Choquet and Triller 2003). As for GlyRs, the insertion of AMPARs happens at the extrasynaptic membrane (Passafaro et al. 2001; Parks et al. 2004; Gerges et al. 2006; Greger and Esteban 2007). NMDARs also have their own life, and they are internalized in response to glutamate or glycine. As for AMPARs, trafficking mechanisms depend on NMDAR subunits. For example, there is a shift in composition during development: the subunit NR2B present early in development is replaced by the NR2A subunit in an activity-dependent manner (reviewed in Collingridge et al. 2004).

In addition to basal turnover, which may reflect the necessary replacement of proteins targeted to degradation, there are neuronal activity phenomena that rely on trafficking of receptors. Such is the case for synaptic plasticity, which increases or decreases receptor number during LTP and LTD, respectively. Actually, simultaneous presynaptic and postsynaptic firing or high-frequency stimulation results in LTP, whereas low-frequency stimulation or uncorrelated firing yields LTD. These modifications are necessary for the development of circuits in the brain and for encoding information in response to experience. For example, brief high-frequency stimulation (tetanus) of the Schaffer-collateral fibers in the hippocampus produces LTP of excitatory synapses and concomitantly LTD of inhibitory ones (Lu et al. 2000). The underlying mechanism includes effects on inhibition through the recruitment of activated calcineurin (CaN) to dephosphorylate GABAARs via the direct binding of the CaN catalytic domain to γ2 subunits (Wang et al. 2003). In addition, potentiation or depression of inhibitory synapses relies on changes in the number of synaptic $GABA_A$Rs (reviewed in Gaiarsa et al. 2002). At excitatory synapses, trafficking of AMPA receptors has a major role in both LTP and LTD. AMPARs are internalized during NMDAR-dependent LTD and inserted into the membrane during NMDAR-dependent LTP, which was found to be associated with the delivery of GFP–GluR1 to spines in organotypic slice cultures (reviewed in Derkach et al. 2007).

Lateral Dynamics and its Regulation

Classically, receptor exchanges during basal turnover or amount modification during plasticity are explained by endocytotic and exocytotic processes. However, a decrease in postsynaptic current due to the loss of receptors from the PSD can be explained not only by the disappearance of receptors from the cell surface but also by their dispersal in the extrasynaptic membrane (see Triller and Choquet 2005). The evolution of imaging technology has allowed the imaging of receptor movements at the single-molecule level. This approach demonstrates that receptors switch at unexpected rates between extrasynaptic and synaptic localizations by lateral diffusion (reviewed in Triller and Choquet 2005; Fig. 2A–B). The continuous exchange of molecules at the PSD can now be generalized to scaffolding molecules as well (Gray et al. 2006).

Lateral Diffusion of Inhibitory Receptors

Initially, the movements of the GlyRs with respect to gephyrin clusters were shown using latex beads (Meier et al. 2001). Most of the time GlyR were freely diffusing but displayed changes in diffusion behavior when associated with submembranous clusters of gephyrin, which caused long confinement periods for the receptors. In these experiments, the size of the probe did not permit investigation of the diffusion at synapses, the synaptic cleft being much smaller. The development of semiconductor quantum dots (QDs) suitable for biological imaging allowed analysis of synaptic GlyR lateral dynamics at the PSD (Dahan et al. 2003). In accordance with the previous study, GlyRs displayed confined diffusion at synapses, whereas their mobility was significantly higher at the extrasynaptic membrane. More importantly, it could be directly demonstrated that GlyRs exchanged continuously between synaptic and extrasynaptic domains, suggesting that the number of synaptic receptors probably depends not only on the capacity of synapses to capture them by interactions with scaffolding molecules but also on lateral diffusion within the plane of the plasma membrane. In a further attempt to understand the regulation of GlyR dynamics, Charrier et al. (2006) analyzed the role of the cytoskeleton. Pharmacological disruption of F-actin and microtubules decreased the amount of GlyR and gephyrin at synapses, increasing the exchange of receptors between the synaptic and extrasynaptic membranes and decreasing their dwell time at synapses. GlyR lateral diffusion was predominantly controlled by microtubules in the extrasynaptic membrane and by actin at synapses. Consequently, the number of synaptic receptors can be rapidly modulated by the cytoskeleton through the regulation of lateral diffusion in the plasma membrane and receptor stabilization at synapses.

The dynamics of GABAR have been investigated by fluorescence recovery after photobleaching (FRAP), an imaging technique that reveals global dynamics. The use of a GABAR subunit bearing the ecliptic pHluorin reporter at the N-terminus allowed the visualization of surface receptors only. It was then found that receptors clustered at synapses had significantly lower mobility than their extrasynaptic counterparts. This reduction in mobility depended on their interaction with gephyrin (Jacob et al. 2005). An electrophysiological study has also allowed the demonstration that GABARs diffuse laterally in the plane of the plasma membrane (Thomas et al. 2005). Importantly, receptor endocytosis occurs exclusively at extrasynaptic sites. Insertions into the plasma membrane are predominantly extrasynaptic and these newly inserted receptors directly access synaptic sites. Therefore, it can be concluded that synaptic $GABA_A$ receptors are directly recruited from their extrasynaptic counterparts by means of lateral diffusion (Bogdanov et al. 2006).

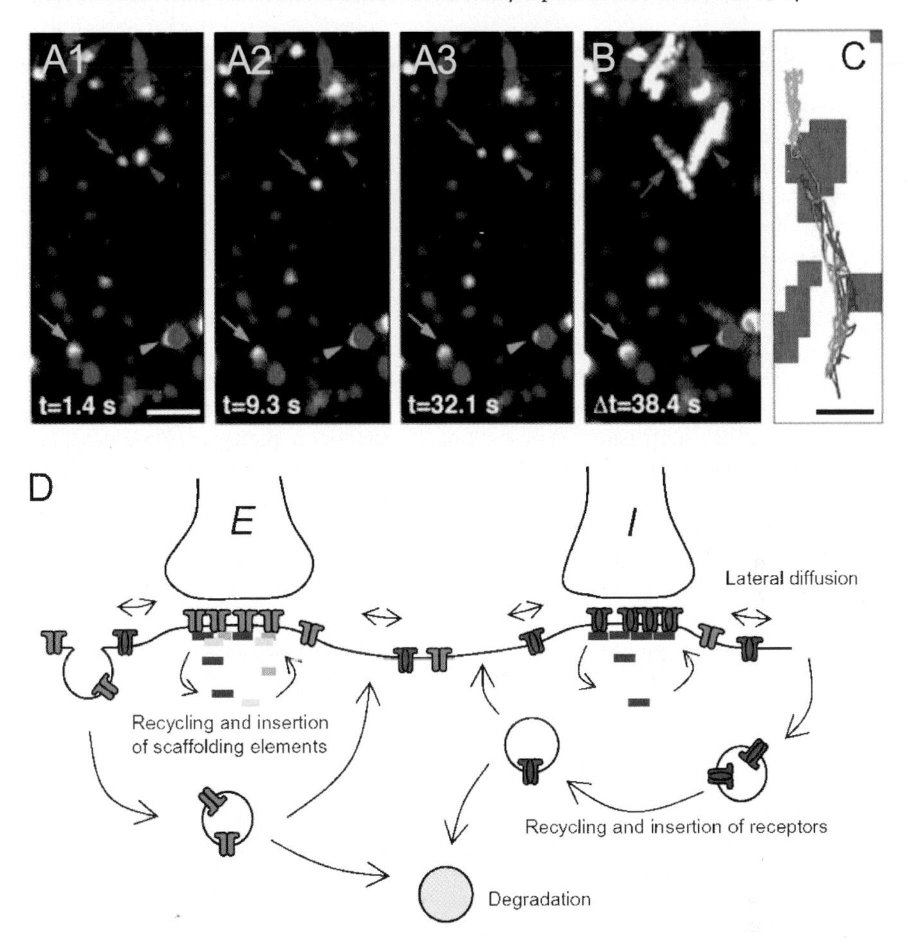

Fig. 2. Receptor trafficking at synapses. **A–C.** Examples of GlyR single particle tracking recordings in cultured spinal cord neurons (10–12 DIV). Active presynaptic terminals were labeled with FM4-64 (red). GlyRs were immunodetected with QDs (GlyR-QDs) (white). **A1–A3.** Sequences of pictures extracted from a recording at the indicated times. **B.** Maximum intensity projections emphasizing the surface area explored by GlyR-QDs during a 38.4-sec recording sequence. Note the differences in the explored surface area and the transitions between synaptic and extrasynaptic compartments. Symbols: *arrows* and *arrowheads* are for identified GlyR-QDs; colors: *green* and *purple*, synaptic and extrasynaptic GlyR-QDs, respectively. Scale bar, 5 μm. **C.** Reconstruction of 18 s GlyR-QD mass center trajectory over the FM4-64-stained synapses (red) images after 1-hour treatment with latrunculin. Scale bar, 1 μm. **D.** Schematic representation of receptor exchanges by lateral diffusion between synaptic, extrasynaptic and intracellular compartments at excitatory (E) and inhibitory (I) contacts. The extrasynaptic receptors are transiently immobilized and enriched at synapses by scaffolding proteins. The extrasynaptic receptors exchange with the intracellular pool, which features receptor synthesis, transport, recycling and degradation. Scaffolding elements at the PSD are in equilibrium with the soluble pools. Each synapse behaves as a receptor donor/acceptor module. (A–C: modified from Charrier et al. 2006, with permission)

Lateral Diffusion of Excitatory Receptors

Single-particle tracking experiments have shown that GluR2-containing AMPARs rapidly alternate between highly mobile and stationary states in a Ca^{2+}-dependent manner (Borgdoff and Choquet 2002). Later studies using smaller probes (single-molecule fluorescence microscopy) to follow diffusion of receptors inside synapses revealed the existence of both immobile and mobile synaptic AMPARs (Tardin et al. 2003; Groc et al. 2004). These results were confirmed by FRAP of pHluorin-tagged AMPARs (Ashby et al. 2004).

Lateral diffusion is probably implicated in the regulation of the number of synaptic receptors during plasticity mechanisms. NMDAR activation leads to the rapid internalization of AMPARs from extrasynaptic sites, which precedes the loss of AMPARs from synapses (Ashby et al. 2004). Bath application of glutamate increased the mobility of synaptic receptors, increasing the proportion of receptors in the area around synapses (Tardin et al. 2003). Furthermore, increasing neuronal activity by bath application of KCl or decreasing it by tetradotoxin (TTX) modified AMPAR but not NMDAR mobility, whereas protein kinase C activation modified both (Groc et al. 2004). Hence a general model is now emerging in which synaptic plasticity involves both regulated exocytosis and endocytosis of AMPARs at extrasynaptic sites and their regulated lateral diffusion into and out of the synapse. Stargazin is likely to play an important role in its regulation since AMPAR lateral dynamics were recently shown to be dependent on the interaction of stargazin with PSD-95 (Bats et al. 2007).

Dynamics of the Scaffold

PSD scaffolding molecules also have faster dynamics than expected. Experiments on young spinal cord neurons without synapses revealed that scaffolds containing a venus::gephyrin chimera displayed lateral movements in a calcium-dependent fashion. Rearrangements within the gephyrin scaffold as well as lateral displacements were also modulated by the cytoskeleton and activity (Hanus et al. 2006).

FRAP has been applied to analyze the turnover of PSD proteins like actin (Star et al. 2002), PSD-95 and PSD-Zip45 (Homer 1c; Okabe et al. 1999, 2001). In the case of PSD-95, more than 20% of GFP-PSD-95 clusters turned over within 24 hours. PSD-95, GKAP, Shank and PSD-Zip45 have a dynamic fraction with time constants of several minutes, PSD-95 being the most stable (Kuriu et al. 2006). As for receptors, the dynamic behavior of scaffolding molecules can be regulated by activity and the cytoskeleton. Dynamics of PSD-95 clusters are inhibited by blockers of excitatory synaptic transmission (Okabe et al. 1999). Acute pharmacological disruption of F-actin eliminates the dynamic fraction of GKAP, Shank and PSD-Zip45, but not PSD-95 (Kuriu et al. 2006). GKAP clusterization at synapses is increased by pharmacological enhancement of neuronal activity, whereas the opposite occurs for Shank and PSD-Zip45. This effect is dependent on F-actin dynamics, as its inhibition prevented the activity-dependent redistribution of all three scaffolds (Kuriu et al. 2006).

The maintenance of stable postsynaptic structures despite the permanent exchange of its components seems to be the rule. An interesting concept that is now emerging is that molecules can belong to different structures. FRAP studies revealed that ProSAP2/Shank3 is continuously redistributed into synaptic structures on time scales of minutes to hours. Importantly, inhibition of protein synthesis or degradation does not affect exchange rates. Therefore, the dynamics of this molecule are dominated by local protein exchange and redistribution (Tsuriel et al. 2006).

Real-time molecular dynamics have mainly been studied in cultured cells, but this field is rapidly evolving and the first report of dynamics of a scaffolding molecule in a living animal was recently published. Gray et al. (2006) studied the dynamics of synaptic PSD-95 tagged with GFP using two-photon microscopy. It was found that PSD-95 clusters are stable for days, but individual PSD-95 molecules turn over rapidly, spending between 20 and 60 minutes at given synapses. Interestingly, retention times measured in vivo are compatible with those measured in cultured neurons (Okabe et al. 2001).

Extrasynaptic Versus Synaptic Receptors

Although evidence from electrophysiological experiments has long since shown the existence of extrasynaptic receptors, they were often taken as a separate pool distinct from synaptic ones. Their physiological role has been limited to activation by spillover of neurotransmitter outside the synaptic cleft during massive release or during transmitter release by neighboring glial cells. The notion that extrasynaptic and synaptic receptors are separate entities was reinforced by the fact that some receptor isoforms have specific subcellular distributions (reviewed in Lüscher and Keller 2004; van Zundert et al. 2004). The ratio between the amounts of synaptic versus extrasynaptic receptors is generally found to be higher by immunocytochemistry, due to the failure of immuno-based methods to detect molecules at low density. Nevertheless, the existence of an extrasynaptic population indistinguishable from the synaptic one is now clear from evidence from trafficking studies. Recently, highly sensitive electron microscopy allowed detection and estimation of AMPAR density at the extrasynaptic membrane (Tanaka et al. 2005). Although the density of extrasynaptic receptors is lower than that of the synaptic ones, extrasynaptic surface receptors are much more numerous than synaptic ones because the extrasynaptic membrane has a larger surface area than the synaptic membrane. Therefore, the extrasynaptic membrane can act as a reserve pool of receptors that can be rapidly mobilized into the synapse in response to activity. In other words, depending on the situation, synapses behave as donor or acceptor for receptors and the pool of extrasynaptic receptors seems to be necessary for regulation of synaptic receptor numbers (3A).

Source of Dynamic Behavior and Membrane Compartmentalization

Molecules are constantly moving, colliding between themselves and bouncing back and forth due to thermal agitation processes. This movement is referred to as Brownian motion and does not have any particular direction. As Albert Einstein clairvoyantly proposed in his PhD thesis (1905), these microscopic movements are translated into diffusion at a macroscopic level. As a consequence, lateral diffusion depends on temperature and density of particles and tends towards homogeneity. Molecules will

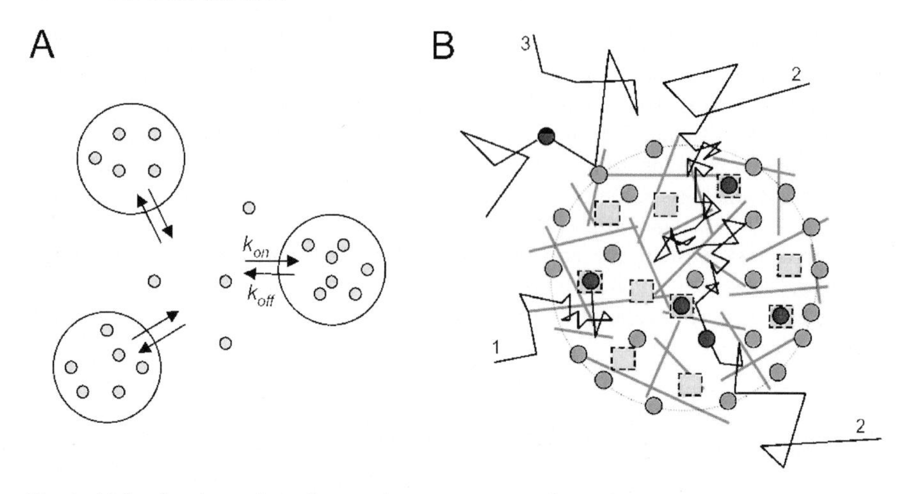

Fig. 3. Molecular dynamic and synaptic structure seen from above. **A.** Receptors (*light blue circles*) enter and exit the PSD with equilibrium constants k_{on} and k_{off}, respectively. **B.** Inside the synaptic area, receptors diffuse (*trajectories 1* and *2*) with reduced mobility in comparison with the extrasynaptic membrane, due to the presence of obstacles such as the underlying cytoskeleton acting as fences (*blue lines*) and transmembrane proteins (*green circles*). They are stabilized at the PSD (*trajectory 1*) by interactions with scaffolding elements (*yellow squares*). Some receptors do not enter the postsynaptic cluster (*trajectory 3*), due to their failure to percolate obstacles and fences at the periphery of the synapse

spread following their concentration gradient to reach a homogeneous distribution. The original mosaic model of biomembranes of Singer and Nicholson (1972) was based on these processes and pointed out the lack of lateral organization of membrane molecules. However, diffusion measurements did not completely fit this model. Lipids, for example, have diffusion coefficients about 100 times lower in biological membranes than in artificial ones, probably due to the presence of proteins. Indeed, the amount of surface covered by proteins is much higher than previously thought (reviewed in Jacobson et al. 2007).

Interestingly, single particle tracking revealed that molecules do not always move with Brownian diffusion, leading to the concept of membrane compartmentalization. Several hypotheses have been proposed to explain the origin of these microdomains, like corralling by submembranous fences, obstacles and variations in membrane fluidity (Kusumi et al. 2005; reviewed in Marguet et al. 2006). Compartmentalization could also arise from patches of membrane with different lipid composition, known as lipid-rafts. The current consensus is that they are rather small (few nanometers) and very dynamic (reviewed in Jacobson et al. 2007). More heterogeneity results from the fact that proteins affect neighboring lipid molecules, creating a ring or "shell" of lipids around them with specific biophysical properties. Furthermore, proteins themselves cause local changes in viscosity, especially in crowded environments (Jacobson et al. 2007).

In summary, thermal agitation can be reduced in various ways. As a consequence, the diffusion of receptors can be slowed down in a given subcellular domain, like synapses, due to local restrictions on diffusion. In fact, about half of the receptors still diffuse at the synaptic membrane and display confined diffusion. Interestingly, the mechanisms causing this reduction in mobility can be regulated. For example, cytoskeleton depolymerization or stabilization has direct consequences for the mobility of receptors (Charrier et al. 2006) and is itself modulated by synaptic activity (reviewed in Dillon and Goda 2005; 3B).

Reconciling Molecular Instability and Permanence of the Synapse

The concept of steady-state structures is central to biology. Any living entity is an example of it, exchanging materials and energy ceaselessly with its environment without losing its identity. Recent improvements in imaging technology allow the application of this concept to molecular assemblies. In particular, the possibility of individualizing molecules is revealing the rules governing the dynamic processes that are hidden by the averaging inherent in bulk experiments. Together with single-channel recordings, these techniques support the emergence of a molecular physiology.

The data reviewed above support the concept of the synapse as a multimolecular assembly at quasi-equilibrium. In other words, the synapse maintains its identity *thanks* to the continuous exchange of its components. The conditions of the exchange (type of molecules and rate) determine the set point of a local equilibrium. Therefore, the molecular composition of synapses is expected to fluctuate around this set point while the quasi-equilibrium state is maintained. Nevertheless, changing the conditions appropriately leads to another set point and, concomitantly, to another local equilibrium. The transition between equilibrium states relies on diffusion of molecules (including receptors and scaffold proteins) in and out of synapses. Thus, plasticity events could be achieved by modulating the flux of molecules that enter and exit the synaptic area.

The rapid exchange between synaptic and extrasynaptic receptors does not exclude the existence of a more stable synaptic pool with a long dwell time. Actually, GlyRs can be split into two populations of different mobility on gephyrin clusters (Ehrensperger et al. 2007). In an electrophysiological approach, surface AMPA receptors were silenced by a photoreactive irreversible antagonist (Adesnik et al. 2005). This study confirmed that the most dynamic movement of AMPARs occurs by lateral movement, and cycling of surface receptors with receptors from internal stores happens exclusively at extrasynaptic somatic sites. On the other hand, the complete replacement of synaptic AMPARs required more than 16 hr.

To analyze the consequences of these dynamic events in more detail, it is necessary to formalize the experimental results into mathematical models that predict synaptic responses. We recently proposed a model to compute the postsynaptic component of synaptic weight based on receptor trafficking (Holcman and Triller 2006). Taking into account the receptor fluxes, it allows the calculation of the amount of receptors bound to the scaffold and its fluctuations. This model permits, for example, determination of the chemical binding constants of receptor-scaffold interactions from data obtained in FRAP experiments.

In cellular energetics, the most "expensive" processes, like protein synthesis and degradation, may serve to maintain the pools of proteins. The fine and fast tuning of local amounts of molecules would rely on "low-cost" mechanisms that take advantage of the energy already present in the system. At the same time, a pathological state could be derived from fluctuations that cannot be compensated by the quasi-equilibrium state. Small and progressive changes in the composition of the synaptic membrane and/or the PSD that modify the exchange of molecules could have dramatic consequences for synaptic function.

References

Adesnik H, Nicoll RA, England PM (2005) Photoinactivation of native AMPA receptors reveals their real-time trafficking. Neuron 48:977–985

Ashby MC, De La Rue SA, Ralph GS, Uney J, Collingridge GL, Henley JM (2004) Removal of AMPA receptors (AMPARs) from synapses is preceded by transient endocytosis of extrasynaptic AMPARs. J Neurosci 24:5172–5176

Bats C, Groc L, Choquet D (2007) The interaction between Stargazin and PSD-95 regulates AMPA receptor surface trafficking. Neuron 53:719–734

Baude A, Nusser Z, Roberts JD, Mulvihill E, McIlhinney RA, Somogyi P (1993) The metabotropic glutamate receptor (mGluR1 alpha) is concentrated at perisynaptic membrane of neuronal subpopulations as detected by immunogold reaction. Neuron 11:771–787

Bedet C, Bruusgaard JC, Vergo S, Groth-Pedersen L, Eimer S, Triller A, Vannier C (2006) Regulation of gephyrin assembly and glycine receptor synaptic stability. J Biol Chem 281: 30046–30056

Bedford FK, Kittler JT, Muller E, Thomas P, Uren JM, Merlo D, Wisden W, Triller A, Smart TG, Moss SJ (2001) GABA(A) receptor cell surface number and subunit stability are regulated by the ubiquitin-like protein Plic-1. Nature Neurosci 4:908–916

Bogdanov Y, Michels G, Armstrong-Gold C, Haydon PG, Lindstrom J, Pangalos M, Moss SJ (2006) Synaptic GABAA receptors are directly recruited from their extrasynaptic counterparts. Embo J 25:4381–4389

Borgdorff AJ, Choquet D (2002) Regulation of AMPA receptor lateral movements. Nature 417:649–653

Charrier C, Ehrensperger MV, Dahan M, Levi S, Triller A (2006) Cytoskeleton regulation of glycine receptor number at synapses and diffusion in the plasma membrane. J Neurosci 26:8502–8511

Chen X, Vinade L, Leapman RD, Petersen JD, Nakagawa T, Phillips TM, Sheng M, Reese TS (2005) Mass of the postsynaptic density and enumeration of three key molecules. Proc Natl Acad Sci USA 102:11551–11556

Cheng D, Hoogenraad CC, Rush J, Ramm E, Schlager MA, Duong DM, Xu P, Wijayawardana SR, Hanfelt J, Nakagawa T, Sheng M, Peng J (2006) Relative and absolute quantification of postsynaptic density proteome isolated from rat forebrain and cerebellum. Mol Cell Proteomics 5:1158–1170

Choquet D, Triller A (2003) The role of receptor diffusion in the organization of the postsynaptic membrane. Nature Rev Neurosci 4:251–265

Clarke NP, Bolam JP (1998) Distribution of glutamate receptor subunits at neurochemically characterized synapses in the entopeduncular nucleus and subthalamic nucleus of the rat. J Comp Neurol 397:403–420

Collingridge GL, Isaac JT, Wang YT (2004) Receptor trafficking and synaptic plasticity. Nature Rev Neurosci 5:952–962

Dahan M, Levi S, Luccardini C, Rostaing P, Riveau B, Triller A (2003) Diffusion dynamics of glycine receptors revealed by single-quantum dot tracking. Science 302:442–445

Derkach VA, Oh MC, Guire ES, Soderling TR (2007) Regulatory mechanisms of AMPA receptors in synaptic plasticity. Nature Rev Neurosci 8:101–113

Dillon C, Goda Y (2005) The actin cytoskeleton: integrating form and function at the synapse. Annu Rev Neurosci 28:25–55

Dugue GP, Dumoulin A, Triller A, Dieudonne S (2005) Target-dependent use of co-released inhibitory transmitters at central synapses. J Neurosci 25:6490–6498
Dumoulin A, Levi S, Riveau B, Gasnier B, Triller A (2000) Formation of mixed glycine and GABAergic synapses in cultured spinal cord neurons. Eur J Neurosci 12:3883–3892
Ehrensperger MV, Hanus C, Vannier C, Triller A, Dahan M (2007) Multiple association states between glycine receptors and gephyrin identified by SPT analysis. Biophys J 92: 3706–3718
Essrich C, Lorez M, Benson JA, Fritschy JM, Luscher B (1998) Postsynaptic clustering of major GABAA receptor subtypes requires the gamma 2 subunit and gephyrin. Nature Neurosci 1:563–571
Feng G, Tintrup H, Kirsch J, Nichol MC, Kuhse J, Betz H, Sanes JR (1998) Dual requirement for gephyrin in glycine receptor clustering and molybdoenzyme activity. Science 282:1321–1324
Fritschy JM, Schweizer C, Brunig I, Luscher B (2003) Pre- and post-synaptic mechanisms regulating the clustering of type A gamma-aminobutyric acid receptors (GABAA receptors). Biochem Soc Trans 31:889–892
Fuhrmann JC, Kins S, Rostaing P, El Far O, Kirsch J, Sheng M, Triller A, Betz H, Kneussel M (2002) Gephyrin interacts with Dynein light chains 1 and 2, components of motor protein complexes. J Neurosci 22:5393–5402
Gaiarsa JL, Caillard O, Ben-Ari Y (2002) Long-term plasticity at GABAergic and glycinergic synapses: mechanisms and functional significance. Trends Neurosci 25:564–570
Gerges NZ, Backos DS, Rupasinghe CN, Spaller MR, Esteban JA (2006) Dual role of the exocyst in AMPA receptor targeting and insertion into the postsynaptic membrane. Embo J 25:1623–1634
Giesemann T, Schwarz G, Nawrotzki R, Berhorster K, Rothkegel M, Schluter K, Schrader N, Schindelin H, Mendel RR, Kirsch J, Jockusch BM (2003) Complex formation between the postsynaptic scaffolding protein gephyrin, profilin, and Mena: a possible link to the microfilament system. J Neurosci 23:8330–8339
Gray NW, Weimer RM, Bureau I, Svoboda K (2006) Rapid redistribution of synaptic PSD-95 in the neocortex in vivo. PLoS Biol 4:e370
Greger IH, Esteban JA (2007) AMPA receptor biogenesis and trafficking. Curr Opin Neurobiol 17:289–297
Groc L, Heine M, Cognet L, Brickley K, Stephenson FA, Lounis B, Choquet D (2004) Differential activity-dependent regulation of the lateral mobilities of AMPA and NMDA receptors. Nat Neurosci 7:695–696
Hanus C, Vannier C, Triller A (2004) Intracellular association of glycine receptor with gephyrin increases its plasma membrane accumulation rate. J Neurosci 24:1119–1128
Hanus C, Ehrensperger MV, Triller A (2006) Activity-dependent movements of postsynaptic scaffolds at inhibitory synapses. J Neurosci 26:4586–4595
Holcman D, Triller A (2006) Modeling synaptic dynamics driven by receptor lateral diffusion. Biophys J 91:2405–2415
Holtmaat A, Wilbrecht L, Knott GW, Welker E, Svoboda K (2006) Experience-dependent and cell-type-specific spine growth in the neocortex. Nature 441:979–983
Jacob TC, Bogdanov YD, Magnus C, Saliba RS, Kittler JT, Haydon PG, Moss SJ (2005) Gephyrin regulates the cell surface dynamics of synaptic GABAA receptors. J Neurosci 25:10469–10478
Jacobson K, Mouritsen OG, Anderson RG (2007) Lipid rafts: at a crossroad between cell biology and physics. Nature Cell Biol 9:7–14
Jordan BA, Fernholz BD, Boussac M, Xu C, Grigorean G, Ziff EB, Neubert TA (2004) Identification and verification of novel rodent postsynaptic density proteins. Mol Cell Proteomics 3:857–871
Kim E, Sheng M (2004) PDZ domain proteins of synapses. Nature Rev Neurosci 5:771–781
Kim EY, Schrader N, Smolinsky B, Bedet C, Vannier C, Schwarz G, Schindelin H (2006) Deciphering the structural framework of glycine receptor anchoring by gephyrin. Embo J 25:1385–1395
Kins S, Betz H, Kirsch J (2000) Collybistin, a newly identified brain-specific GEF, induces submembrane clustering of gephyrin. Nature Neurosci 3:22–29

Kirsch J, Betz H (1995) The postsynaptic localization of the glycine receptor-associated protein gephyrin is regulated by the cytoskeleton. J Neurosci 15:4148–4156

Kirsch J, Wolters I, Triller A, Betz H (1993) Gephyrin antisense oligonucleotides prevent glycine receptor clustering in spinal neurons. Nature 366:745–748

Kittler JT, Moss SJ (2003) Modulation of GABAA receptor activity by phosphorylation and receptor trafficking: implications for the efficacy of synaptic inhibition. Curr Opin Neurobiol 13:341–347

Kittler JT, Rostaing P, Schiavo G, Fritschy JM, Olsen R, Triller A, Moss SJ (2001) The subcellular distribution of GABARAP and its ability to interact with NSF suggest a role for this protein in the intracellular transport of GABA(A) receptors. Mol Cell Neurosci 18:13–25

Kneussel M, Betz H (2000) Clustering of inhibitory neurotransmitter receptors at developing postsynaptic sites: the membrane activation model. Trends Neurosci 23:429–435

Kneussel M, Brandstatter JH, Laube B, Stahl S, Muller U, Betz H (1999) Loss of postsynaptic GABA(A) receptor clustering in gephyrin-deficient mice. J Neurosci 19:9289–9297

Knuesel I, Mastrocola M, Zuellig RA, Bornhauser B, Schaub MC, Fritschy JM (1999) Short communication: altered synaptic clustering of GABAA receptors in mice lacking dystrophin (mdx mice). Eur J Neurosci 11:4457–4462

Kuriu T, Inoue A, Bito H, Sobue K, Okabe S (2006) Differential control of postsynaptic density scaffolds via actin-dependent and -independent mechanisms. J Neurosci 26:7693–7706

Kusumi A, Nakada C, Ritchie K, Murase K, Suzuki K, Murakoshi H, Kasai RS, Kondo J, Fujiwara T (2005) Paradigm shift of the plasma membrane concept from the two-dimensional continuum fluid to the partitioned fluid: high-speed single-molecule tracking of membrane molecules. Annu Rev Biophys Biomol Struct 34:351–378

Legendre P (2001) The glycinergic inhibitory synapse. Cell Mol Life Sci 58:760–93

Levi S, Chesnoy-Marchais D, Sieghart W, Triller A (1999) Synaptic control of glycine and GABA(A) receptors and gephyrin expression in cultured motoneurons. J Neurosci 19:7434–7449

Levi S, Vannier C, Triller A (1998) Strychnine-sensitive stabilization of postsynaptic glycine receptor clusters. J Cell Sci 111 (Pt 3):335–345

Levi S, Grady RM, Henry MD, Campbell KP, Sanes JR, Craig AM (2002) Dystroglycan is selectively associated with inhibitory GABAergic synapses but is dispensable for their differentiation. J Neurosci 22:4274–4285

Levi S, Logan SM, Tovar KR, Craig AM (2004) Gephyrin is critical for glycine receptor clustering but not for the formation of functional GABAergic synapses in hippocampal neurons. J Neurosci 24:207–217

Lu YM, Mansuy IM, Kandel ER, Roder J (2000) Calcineurin-mediated LTD of GABAergic inhibition underlies the increased excitability of CA1 neurons associated with LTP. Neuron 26:197–205

Lujan R, Roberts JD, Shigemoto R, Ohishi H, Somogyi P (1997) Differential plasma membrane distribution of metabotropic glutamate receptors mGluR1 alpha, mGluR2 and mGluR5, relative to neurotransmitter release sites. J Chem Neuroanat 13:219–241

Luscher B, Keller CA (2004) Regulation of GABAA receptor trafficking, channel activity, and functional plasticity of inhibitory synapses. Pharmacol Ther 102:195–221

Mammoto A, Sasaki T, Asakura T, Hotta I, Imamura H, Takahashi K, Matsuura Y, Shirao T, Takai Y (1998) Interactions of drebrin and gephyrin with profilin. Biochem Biophys Res Commun 243:86–89

Marguet D, Lenne PF, Rigneault H, He HT (2006) Dynamics in the plasma membrane: how to combine fluidity and order. Embo J 25:3446–3457

Masugi-Tokita M, Tarusawa E, Watanabe M, Molnar E, Fujimoto K, Shigemoto R (2007) Number and density of AMPA receptors in individual synapses in the rat cerebellum as revealed by SDS-digested freeze-fracture replica labeling. J Neurosci 27:2135–2144

Matsubara A, Laake JH, Davanger S, Usami S, Ottersen OP (1996) Organization of AMPA receptor subunits at a glutamate synapse: a quantitative immunogold analysis of hair cell synapses in the rat organ of Corti. J Neurosci 16:4457–4467

Meier J, Vannier C, Serge A, Triller A, Choquet D (2001) Fast and reversible trapping of surface glycine receptors by gephyrin. Nature Neurosci 4:253–260
Meyer G, Kirsch J, Betz H, Langosch D (1995) Identification of a gephyrin binding motif on the glycine receptor beta subunit. Neuron 15:563–572
Moss SJ, Smart TG (2001) Constructing inhibitory synapses. Nature Rev Neurosci 2:240–250
Nicoll RA, Tomita S, Bredt DS (2006) Auxiliary subunits assist AMPA-type glutamate receptors. Science 311:1253–1256
Nusser Z, Mulvihill E, Streit P, Somogyi P (1994) Subsynaptic segregation of metabotropic and ionotropic glutamate receptors as revealed by immunogold localization. Neuroscience 61:421–427
Okabe S, Kim HD, Miwa A, Kuriu T, Okado H (1999) Continual remodeling of postsynaptic density and its regulation by synaptic activity. Nature Neurosci 2:804–811
Okabe S, Urushido T, Konno D, Okado H, Sobue K (2001) Rapid redistribution of the postsynaptic density protein PSD-Zip45 (Homer 1c) and its differential regulation by NMDA receptors and calcium channels. J Neurosci 21:9561–9571
Park M, Penick EC, Edwards JG, Kauer JA, Ehlers MD (2004) Recycling endosomes supply AMPA receptors for LTP. Science 305:1972–1975
Passafaro M, Piech V, Sheng M (2001) Subunit-specific temporal and spatial patterns of AMPA receptor exocytosis in hippocampal neurons. Nature Neurosci 4:917–926
Peng J, Kim MJ, Cheng D, Duong DM, Gygi SP, Sheng M (2004) Semiquantitative proteomic analysis of rat forebrain postsynaptic density fractions by mass spectrometry. J Biol Chem 279:21003–21011
Petersen JD, Chen X, Vinade L, Dosemeci A, Lisman JE, Reese TS (2003) Distribution of postsynaptic density (PSD)-95 and Ca2+/calmodulin-dependent protein kinase II at the PSD. J Neurosci 23:11270–11278
Racca C, Gardiol A, Triller A (1997) Dendritic and postsynaptic localizations of glycine receptor alpha subunit mRNAs. J Neurosci 17:1691–1700
Ramming M, Kins S, Werner N, Hermann A, Betz H, Kirsch J (2000) Diversity and phylogeny of gephyrin: tissue-specific splice variants, gene structure, and sequence similarities to molybdenum cofactor-synthesizing and cytoskeleton-associated proteins. Proc Natl Acad Sci USA 97:10266–10271
Rasmussen H, Rasmussen T, Triller A, Vannier C (2002) Strychnine-blocked glycine receptor is removed from synapses by a shift in insertion/degradation equilibrium. Mol Cell Neurosci 19:201–215
Rathenberg J, Kittler JT, Moss SJ (2004) Palmitoylation regulates the clustering and cell surface stability of GABAA receptors. Mol Cell Neurosci 26:251–257
Rosenberg M, Meier J, Triller A, Vannier C (2001) Dynamics of glycine receptor insertion in the neuronal plasma membrane. J Neurosci 21:5036–5044
Sabatini DM, Barrow RK, Blackshaw S, Burnett PE, Lai MM, Field ME, Bahr BA, Kirsch J, Betz H, Snyder SH (1999) Interaction of RAFT1 with gephyrin required for rapamycin-sensitive signaling. Science 284:1161–1164
Sheng M, Kim MJ (2002) Postsynaptic signaling and plasticity mechanisms. Science 298:776–780
Sheng M, Hoogenraad CC (2006) The postsynaptic architecture of excitatory synapses: a more quantitative view. Annu Rev Biochem doi:10.1146/annurev.biochem.76.060805.160029
Sola M, Bavro VN, Timmins J, Franz T, Ricard-Blum S, Schoehn G, Ruigrok RW, Paarmann I, Saiyed T, O'Sullivan GA, Schmitt B, Betz H, Weissenhorn W (2004) Structural basis of dynamic glycine receptor clustering by gephyrin. Embo J 23:2510–25199
Somogyi P, Tamas G, Lujan R, Buhl EH (1998) Salient features of synaptic organisation in the cerebral cortex. Brain Res Brain Res Rev 26:113–135
Star EN, Kwiatkowski DJ, Murthy VN (2002) Rapid turnover of actin in dendritic spines and its regulation by activity. Nature Neurosci 5:239–246

Tanaka J, Matsuzaki M, Tarusawa E, Momiyama A, Molnar E, Kasai H, Shigemoto R (2005) Number and density of AMPA receptors in single synapses in immature cerebellum. J Neurosci 25:799–807
Tardin C, Cognet L, Bats C, Lounis B, Choquet D (2003) Direct imaging of lateral movements of AMPA receptors inside synapses. Embo J 22:4656–4665
Thomas P, Mortensen M, Hosie AM, Smart TG (2005) Dynamic mobility of functional GABAA receptors at inhibitory synapses. Nature Neurosci 8:889–897
Triller A, Choquet D (2003) Synaptic structure and diffusion dynamics of synaptic receptors. Biol Cell 95:465–476
Triller A, Choquet D (2005) Surface trafficking of receptors between synaptic and extrasynaptic membranes: and yet they do move! Trends Neurosci 28:133–139
Triller A, Cluzeaud F, Pfeiffer F, Betz H, Korn H (1985) Distribution of glycine receptors at central synapses: an immunoelectron microscopy study. J Cell Biol 101:683–688
Tsuriel S, Geva R, Zamorano P, Dresbach T, Boeckers T, Gundelfinger ED, Garner CC, Ziv NE (2006) Local sharing as a predominant determinant of synaptic matrix molecular dynamics. PLoS Biol 4:e271
van Zundert B, Yoshii A, Constantine-Paton M (2004) Receptor compartmentalization and trafficking at glutamate synapses: a developmental proposal. Trends Neurosci 27:428–437
Vannier C, Triller A (2007) Glycine Receptors: Molecular and Cellular Biology. In: Squire L (ed) New encyclopaedia of neuroscience. Elsevier, Oxford, in press
Wang J, Liu S, Haditsch U, Tu W, Cochrane K, Ahmadian G, Tran L, Paw J, Wang Y, Mansuy I, Salter MM, Lu YM (2003) Interaction of calcineurin and type- A GABA receptor gamma 2 subunits produces long-term depression at CA1 inhibitory synapses. J Neurosci 23:826–836

Cellular Biology of AMPA Receptor Trafficking and Synaptic Plasticity

José A. Esteban[1]

Summary. AMPA-type glutamate receptors are among the most dynamic components of excitatory synapses. Their regulated addition and removal from synapses lead to long-lasting forms of synaptic plasticity, known as long-term potentiation (LTP) and long-term depression (LTD). In addition, AMPA receptors (AMPARs) reach their synaptic targets after a complicated journey involving multiple transport steps through different membrane compartments. This chapter summarizes our current knowledge of the trafficking pathways of AMPARs and their relation to synaptic function and plasticity.

Introduction

Intracellular membrane trafficking is an essential process in all eukaryotic cells, but it is particularly critical at synaptic terminals, where a large number of specific ion channels, scaffolding molecules and multiple signal transduction regulators have to be precisely targeted to ensure proper synaptic function (McGee and Bredt 2003; Ziv and Garner 2004). At the level of the postsynaptic terminal, local membrane trafficking is now appreciated as a major factor controlling synaptic function (Kennedy and Ehlers 2006). In particular, the regulation of neurotransmitter receptor transport and targeting is crucial for the maintenance of synaptic strength and for the activity-dependent changes associated with synaptic plasticity (Collingridge et al. 2004).

Most excitatory transmission in the central nervous system is mediated by two types of glutamate receptors: γ-amino-3-hydroxy-5-methylisoxazole-4-proprionic acid (AMPA) and N-methyl-D-aspartate (NMDA) receptors. These two types of receptors have very different roles in synaptic function (Dingledine et al. 1999; Hollmann and Heinemann 1994). AMPA receptors (AMPARs) mediate most excitatory (depolarizing) currents in conditions of basal neuronal activity, and hence they have a major influence on the strength of the synaptic response. NMDA receptors (NMDARs), on the other hand, remain silent at resting membrane potential (Nowak et al. 1984), but they are crucial for the induction of specific forms of synaptic plasticity, such as long-term potentiation (LTP) and long-term depression (LTD; Bear and Malenka 1994). Although AMPARs and NMDARs reside in the same synapses in most brain regions, they reach their synaptic targets through quite different programs. In the brain, soon after birth, most excitatory synapses in the hippocampus (Durand et al. 1996; Hsia et al. 1998; Petralia et al. 1999) and other brain regions (Feldman et al. 1999; Isaac et al. 1997; Losi et al. 2002; Wu et al. 1996) contain only NMDARs, whereas the prevalence of AMPARs

[1] Department of Pharmacology, University of Michigan Medical School, Ann Arbor, MI 48109-0632, USA, email: estebanj@umich.edu

Selkoe et al.
Synaptic Plasticity and the Mechanism of Alzheimer's Disease
© Springer-Verlag Berlin Heidelberg 2008

increases gradually over the course of postnatal development. In fact, the delivery of AMPARs into synapses is a regulated process that depends on NMDAR activation and underlies some forms of synaptic plasticity in early postnatal development (Zhu et al. 2000) and in mature neurons (Barry and Ziff 2002; Hayashi et al. 2000; Malinow and Malenka 2002; Sheng and Lee 2001; Song and Huganir 2002).

Synaptic plasticity is thought to underlie higher cognitive functions, such as learning and memory (Bliss and Collingridge 1993; Chen and Tonegawa 1997; Elgersma and Silva 1999; Hebb 1949; Martin et al. 2000), and is also critical for neural development (Cline 1998; Constantine-Paton 1990; Katz and Shatz 1996). Thus, it is not surprising that alterations in synaptic plasticity have been implicated in the pathology of several neurological disorders, including Alzheimer's disease (Esteban 2004; Rowan et al. 2003; Turner et al. 2003), schizophrenia (Konradi and Heckers 2003; Stephan et al. 2006), Down's syndrome (Galdzicki and Siarey 2003) and other forms of mental retardation (Newey et al. 2005). Consequently, there is considerable interest in understanding the underlying mechanisms of synaptic plasticity, among which the regulation of AMPAR trafficking plays a prominent role.

This review will summarize our current knowledge of the membrane trafficking pathways that steer AMPARs from their biosynthesis at the endoplasmic reticulum (ER) to their destination at excitatory synapses, with special emphasis on the regulatory steps that contribute to synaptic plasticity. Most of the experimental observations that are the basis for this chapter have been obtained from hippocampal principal neurons, although it is expected that most of the principles described here will be applicable to the regulation of AMPAR trafficking in multiple brain regions.

AMPA Receptor Assembly and Exit from the Endoplasmic Reticulum

AMPARs are hetero-tetramers (Rosenmund et al. 1998), composed of different combinations of GluR1, GluR2, GluR3, and GluR4 subunits (Hollmann and Heinemann 1994). In the mature hippocampus, most AMPARs are composed of GluR1–GluR2 or GluR2–GluR3 combinations (Wenthold et al., 1996), whereas GluR4-containing AMPARs are expressed mainly in early postnatal development (Zhu et al. 2000). These oligomeric combinations are formed in the ER, possibly assembling as dimers of dimers (Tichelaar et al. 2004) via interactions between the luminal, N-terminal domains of the subunits (Greger et al. 2007; Kuusinen et al. 1999; Leuschner and Hoch 1999). After assembly, exit from the ER is tightly regulated by quality control mechanisms that monitor the competency of newly synthesized receptors for ligand binding and gating (Fleck 2006).

Interestingly, AMPAR trafficking through the ER is subunit-specific. Thus, GluR1–GluR2 hetero-oligomers exit the ER rapidly and traffic to the Golgi compartment, where they become fully glycosylated (Greger et al. 2002). In contrast, GluR2–GluR3 heteromers take much longer to exit (i.e., are retained longer in) the ER. In fact, a fraction of the GluR2 subunits seems to remain unassembled within the ER, in a manner that depends on the presence of an edited arginine residue (R607) at the channel pore region (Greger et al. 2002, 2003). These immature AMPAR subunits appear to associate with molecular chaperones residing at the ER (Fukata et al. 2005; Greger et al. 2002). Interactions with cytosolic proteins also seem to control trafficking through the ER. For example, the GluR2 C-terminus has a PDZ consensus motif (-SVKI) that interacts with the PDZ domain-containing protein PICK1 (Dev et al. 1999; Perez et al. 2001; Xia et al. 1999). This interaction is required for GluR2's exit from the ER (Greger et al. 2002).

Additionally, export of AMPARs from the ER and surface expression are facilitated by direct interaction with a family of "Transmembrane AMPAR regulatory proteins" (TARPs; Vandenberghe et al. 2005; Ziff 2007). In fact, TARPs may well be considered auxiliary subunits of AMPARs (Fukata et al. 2005), which assist in their proper folding and affect channel kinetics (Bedoukian et al. 2006; Tomita et al. 2005; Turetsky et al. 2005).

AMPA Receptor Trafficking Along the Microtubular Cytoskeleton in Dendrites

Although the dendritic synthesis of AMPARs has been recently reported (Ju et al. 2004), most receptors are likely synthesized in the neuronal body. Therefore, newly synthesized receptors will have to travel long distances from their point of biosynthesis to their final synaptic targets. The long-range dendritic transport of AMPARs probably depends on the microtubular cytoskeleton that runs along dendritic shafts. The transport of membrane organelles on microtubule tracks is an active process powered mainly by motor proteins of the kinesin and dynein superfamilies (Goldstein and Yang 2000). Therefore, membrane compartments bearing AMPARs are likely to be recognized and transported by some of these motor proteins. The molecular mechanisms underlying these processes are still being elucidated.

The PDZ domain-containing protein Glutamate Receptor Interacting Protein 1 (GRIP1) interacts directly with the heavy chain of conventional kinesin, as revealed by yeast two-hybrid screening (Setou et al. 2002). GRIP binds to the C-terminal PDZ motif of GluR2 and GluR3 (Dong et al. 1997; Srivastava et al. 1998), and hence may serve as the link between AMPARs and microtubular motor proteins. In fact, the ternary complex formed by GluR2, GRIP1, and kinesin can be immunoprecipitated from brain lysates, and dominant-negative versions of kinesin reduce the presence of AMPAR at synapses (Setou et al. 2002).

AMPARs have also been shown to associate with a different neuron-specific kinesin motor, KIF1 (Shin et al. 2003). In this case, the adaptor molecule seems to be liprin-α which interacts with GluR2–GRIP (Wyszynski et al. 2002) and with KIF1 (Shin et al. 2003). Another member of the liprin-α–AMPAR–GRIP complex is GIT1, which is also involved in AMPAR trafficking (Ko et al. 2003). Therefore, it seems that the GRIP1–AMPAR complex can be transported along dendrites by more than one type of kinesin motor.

In addition to this microtubular-dependent transport, the export of AMPARs from the cell body into the dendritic surface is powered by a specific actin-based motor protein, myosin Vb (Lise et al. 2006). Interestingly, this transport system is specific for the GluR1 subunit and requires the small GTPase Rab11, possibly acting as a linker between the motor protein and its membrane cargo. From these combined studies, it seems likely that multiple associations between AMPARs and cytoskeletal motor proteins will be discovered in the future, possibly mediated by specific scaffolding molecules.

Actin-Dependent Trafficking in Spines

Most excitatory synapses in the adult brain occur on small dendritic protuberances called spines (Hering and Sheng 2001). Dendritic spines lack microtubular cytoskeleton, but they are rich in highly motile actin filaments (Fischer et al. 1998). Therefore, at some point, AMPAR-containing organelles, trafficking along microtubular tracks, must be transferred to the actin-based cytoskeleton for their final delivery into synapses. The importance of the actin cytoskeleton for local AMPAR trafficking is underscored by the observation that pharmacological depolymerization of actin filaments leads to the removal of AMPARs from dendritic spines (Allison et al. 1998) and from synapses (Kim and Lisman 1999).

The molecular mechanisms that may mediate the actin-based movement of AMPARs are largely unknown. Nevertheless, AMPARs can be linked to the actin cytoskeleton through several scaffolding proteins, such as 4.1N (Shen et al. 2000) and RIL (Schulz et al. 2004). The different members of the protein 4.1 family are known to couple the spectrin–actin cytoskeleton to different membrane-associated proteins (Hoover and Bryant 2000). In particular, the neuronal isoform 4.1N interacts directly with GluR1 (Shen et al. 2000) and GluR4 (Coleman et al. 2003) through the juxtamembrane region of their cytoplasmic C-terminal tails. The other potential actin linker for AMPARs, RIL, is a multi-functional protein that interacts with an internal region of the GluR1 C-terminus through its LIM domain and with α-actinin through its PDZ domain. Interestingly, only AMPAR subunits with long C-tails (GluR1 and GluR4) have been shown so far to couple with the actin cytoskeleton. Since these long-tail subunits are the ones involved in regulated (activity-dependent) trafficking at the synapse (Malinow et al. 2000), it is tempting to speculate that actin-dependent transport may be particularly critical for AMPAR trafficking during synaptic plasticity.

The transport of AMPARs along the spine actin cytoskeleton is likely to be bidirectional, since AMPARs are known to move in and out of synapses in a very dynamic manner. This expectation has been recently confirmed by the identification of an actin-based motor protein, myosin VI, as a mediator of the endocytic removal of AMPARs from synapses (Osterweil et al. 2005). Myosin VI interacts with the GluR1-binding protein SAP97 (Wu et al. 2002), providing a mechanistic link between AMPARs (again through a long-tail subunit) and the motor protein that drives their internalization. Undoubtedly, further studies will be required to unravel what is likely to be a network of interactions mediating the transport of AMPARs along the actin cytoskeleton in synaptic terminals.

TARPs and AMPA Receptor Surface Trafficking

TARPs are the only known transmembrane proteins found to be associated with AMPARs. The first TARP to be identified was stargazin, which was found as a spontaneous mutation in the *stargazer* mouse (Letts et al. 1998) and is critically required for cell surface expression of AMPARs in cerebellar granule cells (Chen et al. 2000). By sequence and structural homology, stargazin belongs to a large group of proteins that includes γ-subunits of Ca^{2+} channels and the claudin family of cell-adhesion molecules.

Nevertheless, only five of these proteins have been described to bind AMPARs and affect their trafficking: stargazin, γ-3, γ-4, γ-8 (Tomita et al., 2003) and, more recently, γ-7 (Kato et al. 2007). Therefore, these are the proteins collectively known as TARPs. Interestingly, different TARPs display specific expression patterns in brain that are to some extent complementary (Tomita et al. 2003).

TARPs associate with AMPARs early in their biosynthetic pathway, as mentioned above, and are able to combine with all AMPAR populations irrespective of their subunit composition (Tomita et al. 2003). The most striking property of TARPs is their critical role for the expression of AMPARs at the extrasynaptic neuronal surface. Genetic ablation of stargazin, the TARP member most abundantly expressed in cerebellum, results in a virtual depletion of AMPARs from the extrasynaptic surface in granule cells (Chen et al. 2000). Similarly, removal of γ-8, a TARP member that is almost exclusively expressed in hippocampus, precludes AMPAR surface expression in hippocampal pyramidal neurons (Rouach et al. 2005). Interestingly, TARPs seem to be a limiting factor for AMPAR cell surface delivery, since overexpression of the appropriate neuron-specific TARP leads to a marked increase in the number of AMPARs expressed on the plasma membrane (Chen et al. 2000; Rouach et al. 2005). The role of these extrasynaptic surface receptors is still debated, although morphological evidence indicates that they are highly mobile and can reach the postsynaptic membrane through lateral diffusion (Borgdorff and Choquet 2002; Choquet and Triller 2003; Groc et al. 2004; Tardin et al. 2003).

TARPs also participate in the trafficking of AMPARs into the synaptic membrane. TARPs contain a PDZ consensus sequence at the C-terminus, which can bind the PDZ domain of MAGUK (membrane-associated guanylate kinase) proteins, such as PSD95 and PSD93 (Chen et al. 2000). MAGUKs are synaptic scaffolding molecules that have been shown to be critical regulators of AMPAR delivery and/or stabilization at synapses (El-Husseini et al. 2000; El-Husseini Ael et al. 2002; Elias et al. 2006; Schluter et al. 2006). Therefore, TARPs are thought to be the molecular linkers between AMPARs and MAGUKs. In particular, the association between TARPs and MAGUKs has been recently shown to be critical to retain AMPARs at synapses. Thus, impairment of the PDZ interaction between stargazin (TARP) and PSD95 (MAGUK) leads to increased receptor diffusion out of the synaptic membrane (Bats et al. 2007). Therefore, a major function of the TARP-MAGUK interaction appears to be the stabilization/anchoring of AMPARs at synapses.

The dual role of TARPs in extrasynaptic surface expression and in receptor stabilization at synapses has led to the hypothesis that AMPAR synaptic delivery occurs in two steps: insertion in the extrasynaptic surface followed by lateral diffusion and synaptic trapping. Indeed, there are morphological (Passafaro et al. 2001) and electrophysiological (Adesnik et al. 2005) observations supporting this scenario. However, there are also indications that extrasynaptic surface receptors are not a necessary source for synaptic delivery. For example, genetic ablation of the hippocampal TARP (γ-8) produced a virtual depletion of extrasynaptic AMPARs, with only a modest effect on the accumulation of AMPARs at synapses (Rouach et al. 2005). Conversely, TARP overexpression produces a massive increase in extrasynaptic AMPARs without any detectable effect on AMPAR-mediated synaptic transmission (Rouach et al. 2005; Schnell et al. 2002). Therefore, the actual contribution of surface diffusion for the synaptic delivery of AMPARs remains to be determined. It is also worth keeping in mind that this contribution may vary for different synapse types and developmental stages.

Subunit-Specificity for Constitutive and Regulated Synaptic Delivery of AMPARs

The final steps in the synaptic trafficking of AMPARs depend on their subunit composition, and specifically, on *cis*-signals contained within their cytosolic carboxy termini (Passafaro et al. 2001; Shi et al. 2001). In hippocampus, hetero-tetramers formed by GluR1/GluR2 and GluR2/GluR3 subunits, together with a smaller contribution from GluR1 homomers, represent the most common combinations in excitatory synapses (Wenthold et al. 1996). Based on experiments expressing recombinant AMPAR subunits in hippocampal neurons, it has been shown that GluR2/GluR3 hetero-tetramers continuously cycle in and out of synapses in a manner largely independent from synaptic activity (Passafaro et al. 2001; Shi et al. 2001). This process (constitutive pathway) preserves the total number of receptors at synapses; therefore, it has been proposed to help maintain synaptic strength in the face of protein turnover (Malinow et al. 2000). This constitutive cycling is very fast (half-time of minutes) and requires a direct interaction between GluR2 and NSF (Nishimune et al. 1998). The precise role of NSF in this trafficking pathway is not fully understood, but it may involve NSF-assisted dissociation of GluR2 from the PDZ domain protein PICK1, allowing AMPARs to cycle back into synapses (Hanley et al. 2002). The continuous synaptic cycling of AMPARs also requires the molecular chaperon Hsp90 (Gerges et al., 2004b), although the mechanistic link between AMPARs and Hsp90 has not been elucidated yet.

In contrast with this constitutive trafficking, AMPARs containing GluR1 (Hayashi et al. 2000), GluR2-long (Kolleker et al. 2003; a splice variant of GluR2; Kohler et al. 1994) or GluR4 (Zhu et al. 2000) subunits are added into synapses in an activity-dependent manner during synaptic plasticity. This regulated pathway is triggered transiently upon induction of LTP and results in a net increase in the number of AMPARs present at synapses (Malinow et al. 2000). The synaptic delivery of GluR1 is also regulated by physiological stimulation in living animals, as it has been reported for neocortical neurons upon sensory stimulation (Takahashi et al. 2003) and in the lateral amygdala after fear conditioning (Rumpel et al. 2005). The subunit composition of the endogenous AMPARs that participate in regulated synaptic delivery has been more difficult to establish. Thus, both GluR2-lacking receptors (presumably GluR1 homomers; Plant et al., 2006) and GluR2-containing receptors (presumably GluR1/GluR2 heteromers; Adesnik and Nicoll 2007; Bagal et al. 2005) have been proposed to be rapidly inserted into synapses upon NMDA receptor activation in hippocampal slices. Although the details remain to be clarified, the importance of subunit composition for the regulation of synaptic delivery is well established and has been recently corroborated by in vivo studies that demonstrated that sensory stimulation (Clem and Barth 2006) or deprivation (Goel et al. 2006), as well as cocaine administration (Bellone and Luscher 2006), can alter the prevalence of AMPARs with different subunit assemblies at synapses.

The activity-dependent synaptic delivery of AMPARs is regulated by several protein kinases, such as CaMKII (reviewed in Lisman and Zhabotinsky 2001), PKA (Ehlers 2000; Esteban et al. 2003; Gao et al. 2006; Gomes et al. 2004; Man et al. 2007), PKC (Boehm et al. 2006; Gomes et al. 2007; Ling et al. 2006) and PI3K (Man et al. 2003). Interestingly, the signaling cascades controlling the delivery of AMPARs to synapses, as well as the AMPAR subunits involved, change during development. Thus, early in postnatal development of the hippocampus, the regulated delivery of AMPARs involves GluR4-containing receptors (Zhu et al. 2000), and PKA-mediated phosphorylation of this subunit triggers receptor delivery (Esteban et al. 2003). Around the second postnatal week, LTP is mostly mediated by the synaptic delivery of GluR2-long (Kolleker et al. 2003). Then, later in development, the regulated addition of AMPARs requires both GluR1 phosphorylation by PKA and CaMKII activation (Esteban et al. 2003). These developmental changes in the regulation of AMPAR synaptic delivery fit very well with the switch in signaling cascades that is required for LTP induction at different postnatal ages (Yasuda et al. 2003).

Local Intracellular Trafficking of AMPARs. Role of Rab Proteins and the Exocyst

Rapid exocytic events can mediate the delivery of AMPARs into synapses (Kopec et al. 2006; Lu et al. 2001; Luscher et al. 1999; Pickard et al. 2001). In this sense, it may come as a surprise that very little is known about the subcellular organization of the membrane trafficking machinery that mediates AMPAR synaptic delivery. This picture has started to change recently, with the identification of local endosomal compartments in close proximity to synapses, or even within dendritic spines, that mediate the delivery of AMPARs into the synaptic membrane (Gerges et al. 2004a; Park et al. 2004, 2006). These new reports are starting to offer a glimpse of the complexity of the membrane trafficking machinery operating at postsynaptic terminals and how it may relate to the subunit-specific synaptic delivery of AMPARs.

Most intracellular membrane sorting in eukaryotic cells is governed by small GTPases of the Rab family (Zerial and McBride 2001). Therefore, the identification of specific Rab proteins involved in AMPAR trafficking may give us some clues as to how the intracellular sorting and synaptic targeting of AMPARs is organized in neurons. It was recently proposed that recycling endosomes driven by the small GTPase Rab11 mediate the activity-dependent delivery of GluR1-containing AMPARs into synapses (Park et al. 2004). In addition, Rab8, which controls *trans*-Golgi network trafficking (Huber et al. 1993) and a separate endosomal population (Hattula et al. 2006), is also required for GluR1 synaptic insertion and LTP (Gerges et al. 2004a). Therefore, it seems that the activity-dependent delivery of AMPARs involves a relay of at least two distinct membrane compartments, whose sorting is controlled by Rab11 and Rab8, possibly acting in separate trafficking steps. Rab11-containing endosomes have recently been localized at the base of dendritic spines (Park et al. 2006), whereas ultrastructural studies have detected Rab8 in close proximity to the postsynaptic membrane (Gerges et al. 2004a). According to these morphological observations, we propose a model in which AMPARs enter spines through Rab11-dependent endosomes. Subsequently, an additional endosomal population, controlled by Rab8, would drive their insertion into the synaptic membrane (see model in Fig. 1).

As mentioned above, in addition to their activity-dependent synaptic delivery, AMPARs are engaged in constitutive trafficking in and out of synapses. This con-

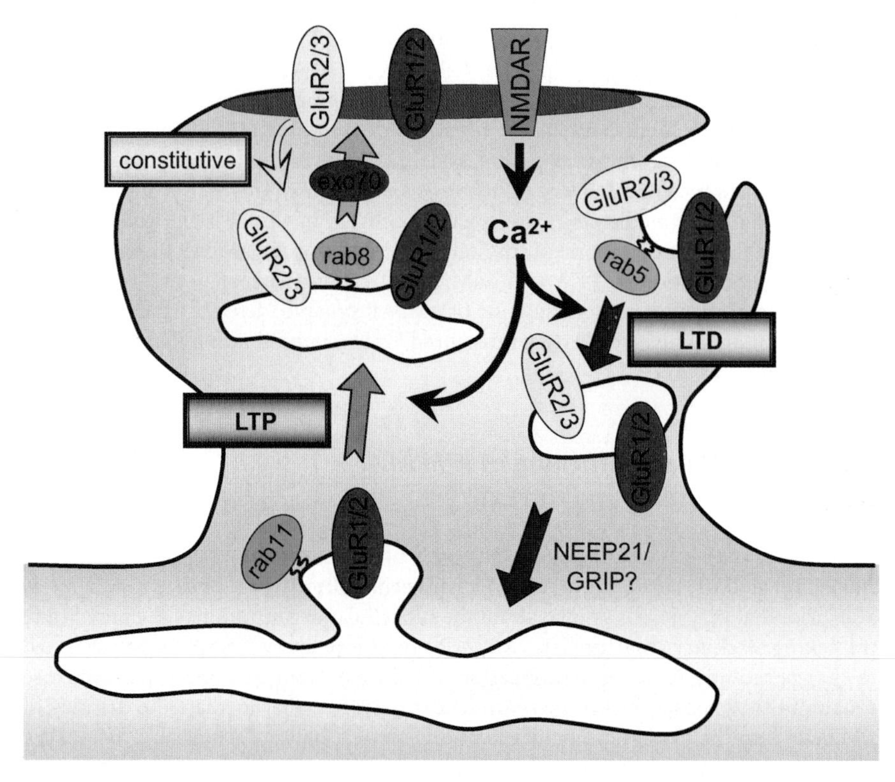

Fig. 1. Schematic model for the endosomal membrane trafficking machinery operating at postsynaptic terminals. The activity-dependent entry of GluR1-containing AMPARs into spines is controlled by Rab11 upon LTP induction. Once within the spine, both GluR1–GluR2 and GluR2–GluR3 AMPARs are driven into synapses in an exocytic process controlled by Rab8 and the exocyst subunit Exo70. In addition, GluR2–GluR3 receptors are engaged in constitutive cycling in and out of the postsynaptic membrane. The activity-dependent internalization of AMPARs is mediated by Rab5, acting on the lateral (extrasynaptic) membrane within the spine. Re-entry of internalized receptors into the Rab11–Rab8 delivery circuit may require the participation of NEEP21 and GRIP

tinuous cycling is known to involve endocytic and exocytic events (Luscher et al. 1999). However, very little is known about the intracellular machinery that controls this process. It has been shown that Rab proteins typically associated with recycling endosomes, such as Rab4 and Rab11, do not participate in constitutive AMPAR synaptic cycling. In contrast, Rab8 appears to be critically required (Gerges et al. 2004a). Since Rab8 is also involved in activity-dependent trafficking (see above), these results indicate that there is a partial overlap between the endosomal machinery mediating constitutive and regulated delivery of AMPARs at synapses (see model in Fig. 1). The endocytic arm of this continuous cycling of receptors is even less characterized. The prototypic Rab protein for endocytosis, Rab5 (Bucci et al. 1992), does not participate in constitutive AMPAR internalization (Brown et al. 2005). Dynamin was shown to be required for this process (Luscher et al. 1999), but the role of clathrin has not been directly tested yet. Obviously, more work will be required to elucidate the cellular basis of this very dynamic aspect of the intracellular trafficking of AMPARs.

The final step in the intracellular trafficking of AMPARs involves their functional insertion and stabilization at the postsynaptic membrane. As mentioned before, several members of the MAGUK family of scaffolding proteins are critical factors in the synaptic targeting of AMPARs (Elias et al. 2006). Interestingly, these synaptic scaffolding molecules associate with the exocyst (Riefler et al. 2003; Sans et al. 2003), a known effector of Rab-dependent exocytic trafficking (Guo et al. 1999; Novick et al. 2006). Therefore, the exocyst may act as a link between incoming AMPAR-containing vesicles and the synaptic scaffold. In agreement with this scenario, it has recently been shown that the exocyst acts within the dendritic spine to mediate the insertion of AMPARs into the postsynaptic membrane (Gerges et al. 2006). In particular, interference with the Exo70 subunit of the exocyst leads to the accumulation of AMPARs within the postsynaptic density, before fusion with the synaptic membrane (Gerges et al. 2006). This observation suggests that AMPAR synaptic insertion from intracellular compartments occurs directly at the level of the postsynaptic density (see model in Fig. 1).

Activity-Dependent Internalization and Sorting of AMPARs

Synaptic AMPARs are internalized in an activity-dependent manner, leading to LTD. This process requires clathrin-mediated endocytosis (Carroll et al. 1999; Lee et al. 2002; Man et al. 2000; see also review in Carroll et al. 2001). Interestingly, and in contrast with constitutive endocytosis, the small GTPase Rab5 drives the regulated internalization of AMPARs during LTD (Brown et al. 2005; see model in Fig. 1). In fact, Rab5 is rapidly and transiently activated upon NMDAR activation during LTD induction (Brown et al. 2005). Therefore, these results suggest that constitutive and regulated AMPAR internalization may engage different components of the endocytic machinery.

In contrast to the subunit-specific rules for AMPAR delivery, the contribution of different receptor populations to activity-dependent removal still remains controversial. Hippocampal neurons lacking both GluR2 and GluR3 subunits display normal LTD, suggesting that GluR1 removal contributes to synaptic depression (Meng et al. 2003). On the other hand, GluR2 subunits are removed during LTD in hippocampal neurons (Seidenman et al. 2003) and cerebellar LTD requires PKC phosphorylation of GluR2 (Chung et al. 2003). Therefore, both GluR1- and GluR2-containing receptors seem to participate in the synaptic trafficking associated with LTD. In fact, most experimental evidence is compatible with an initial indiscriminate internalization of all AMPAR populations upon LTD induction. However, it is increasingly appreciated that AMPARs undergo complicated intracellular sorting and recycling events after synaptic removal, which may involve significant subunit specificity (Lee et al. 2004).

The molecular mechanisms that organize post-endocytic sorting of AMPARs and potential reinsertion into synaptic and/or extra-synaptic membranes are still far from clear. Nevertheless, the balance between GRIP/ABP and PICK1 interactions with GluR2 after PKC phosphorylation seems to be a critical factor (Hanley 2006; Kim et al. 2001; Perez et al. 2001). In hippocampal and parallel fiber-Purkinje cell synapses, PICK1 appears to drive the synaptic removal of phosphorylated GluR2 receptors (Chung et al.

2003; Kim et al. 2001; Steinberg et al. 2006). This role is facilitated by the calcium-dependent interactions between GluR2 and PICK1 (Hanley and Henley 2005). Subsequently, a fraction of these internalized GluR2 subunits recycles back into synaptic sites, in a process probably mediated by direct GRIP/ABP-PICK1 interactions (Lu and Ziff 2005) and NSF-mediated dissociation of the GluR2-PICK1 complex (Hanley et al. 2002). The connections between these AMPAR binding proteins and the intracellular membrane trafficking machinery are still being elucidated, but it has been recently proposed that the return of AMPARs to synaptic sites may be mediated by phosphorylation-regulated interactions between GRIP/ABP and the endosomal protein NEEP21 (Kulangara et al. 2007; Steiner et al. 2005; see model in Fig. 1).

Conclusions

The field of AMPAR trafficking is advancing at a fast pace. New proteins interacting with AMPARs or with the AMPAR trafficking machinery are constantly being identified. These new investigations are uncovering an intricate choreography, in which AMPARs are assembled, sorted and targeted throughout the neuronal secretory pathway. We are starting to identify the core cellular machinery that transports AMPARs, as well as the regulatory molecules that orchestrate their dynamic behavior close to the synapse, where bidirectional AMPAR movement results in long-lasting changes in synaptic strength. These are exciting times, when the fields of AMPAR trafficking and synaptic plasticity are starting to be integrated within the realm of cellular biology.

References

Adesnik H, Nicoll RA (2007) Conservation of glutamate receptor 2-containing AMPA receptors during long-term potentiation. J Neurosci 27; 4598–4602.

Adesnik H, Nicoll RA, England PM (2005) Photoinactivation of native AMPA receptors reveals their real-time trafficking. Neuron 48:977–985.

Allison DW, Gelfand VI, Spector I, Craig AM (1998) Role of actin in anchoring postsynaptic receptors in cultured hippocampal neurons: differential attachment of NMDA versus AMPA receptors. J Neurosci 18:2423–2436.

Bagal AA, Kao JP, Tang CM, Thompson SM (2005) Long-term potentiation of exogenous glutamate responses at single dendritic spines. Proc Natl Acad Sci USA 102: 14434–14439.

Barry MF, Ziff EB (2002) Receptor trafficking and the plasticity of excitatory synapses. Curr Opin Neurobiol 12:279–286.

Bats C, Groc L, Choquet D (2007) The interaction between Stargazin and PSD-95 regulates AMPA receptor surface trafficking. Neuron 53: 719–734.

Bear MF, Malenka RC (1994) Synaptic plasticity: LTP and LTD. Curr Opin Neurobiol 4:389–399.

Bedoukian MA, Weeks AM, Partin KM (2006) Different domains of the AMPA receptor direct stargazin-mediated trafficking and stargazin-mediated modulation of kinetics. J Biol Chem 281:23908–23921.

Bellone C, Luscher C (2006) Cocaine triggered AMPA receptor redistribution is reversed in vivo by mGluR-dependent long-term depression. Nature Neurosci 9:636–641.

Bliss TV, Collingridge GL (1993) A synaptic model of memory: long-term potentiation in the hippocampus. Nature 361:31–39.

Boehm J, Kang MG, Johnson RC, Esteban J, Huganir RL, Malinow R (2006) Synaptic incorporation of AMPA receptors during LTP is controlled by a PKC phosphorylation site on GluR1. Neuron 51:213–225.

Borgdorff AJ, Choquet D (2002) Regulation of AMPA receptor lateral movements. Nature 417:649–653.

Brown TC, Tran IC, Backos DS, Esteban JA (2005). NMDA receptor-dependent activation of the small GTPase Rab5 drives the removal of synaptic AMPA receptors during hippocampal LTD. Neuron 45:81–94.

Bucci C, Parton RG, Mather IH, Stunnenberg H, Simons K, Hoflack B, Zerial M (1992) The small GTPase rab5 functions as a regulatory factor in the early endocytic pathway. Cell 70:715–728.

Carroll RC, Beattie EC, Xia H, Luscher C, Altschuler Y, Nicoll RA, Malenka RC, von Zastrow M (1999) Dynamin-dependent endocytosis of ionotropic glutamate receptors. Proc Natl Acad Sci USA 96:14112–14117.

Carroll RC, Beattie EC, von Zastrow M, Malenka RC (2001) Role of AMPA receptor endocytosis in synaptic plasticity. Nature Rev Neurosci 2:315–324.

Chen C, Tonegawa S (1997) Molecular genetic analysis of synaptic plasticity, activity-dependent neural development, learning, and memory in the mammalian brain. Annu Rev Neurosci 20:157–184.

Chen L, Chetkovich DM, Petralia RS, Sweeney NT, Kawasaki Y, Wenthold RJ, Bredt DS, Nicoll RA (2000) Stargazin regulates synaptic targeting of AMPA receptors by two distinct mechanisms. Nature 408:936–943.

Choquet D, Triller A (2003) The role of receptor diffusion in the organization of the postsynaptic membrane. Nature Rev Neurosci 4:251–265.

Chung HJ, Steinberg JP, Huganir RL, Linden DJ (2003). Requirement of AMPA receptor GluR2 phosphorylation for cerebellar long-term depression. Science 300:1751–1755.

Clem RL, Barth A (2006) Pathway-specific trafficking of native AMPARs by in vivo experience. Neuron 49:663–670.

Cline HT (1998) Topographic maps: developing roles of synaptic plasticity. Curr Biol 8:R836–839.

Coleman SK, Cai C, Mottershead DG, Haapalahti JP, Keinanen K (2003) Surface expression of GluR-D AMPA receptor is dependent on an interaction between its C-terminal domain and a 4.1 protein. J Neurosci 23:798–806.

Collingridge GL. Isaac JT, Wang YT (2004) Receptor trafficking and synaptic plasticity. Nature Rev Neurosci 5:952–962.

Constantine-Paton M (1990) NMDA receptor as a mediator of activity-dependent synaptogenesis in the developing brain. Cold Spring Harb Symp Quant Biol 55:431–443.

Dev KK, Nishimune A, Henley JM, Nakanishi S (1999) The protein kinase C alpha binding protein PICK1 interacts with short but not long form alternative splice variants of AMPA receptor subunits. Neuropharmacology 38:635–644.

Dingledine R, Borges K, Bowie D, Traynelis SF (1999) The glutamate receptor ion channels. Pharmacol Rev 51:7–61.

Dong H, O'Brien RJ, Fung ET, Lanahan AA, Worley PF, Huganir RL (1997) GRIP: a synaptic PDZ domain-containing protein that interacts with AMPA receptors. Nature 386:279–284.

Durand GM, Kovalchuk Y, Konnerth A (1996) Long-term potentiation and functional synapse induction in developing hippocampus. Nature 381:71–75.

Ehlers MD (2000) Reinsertion or degradation of AMPA receptors determined by activity-dependent endocytic sorting. Neuron 28:511–525.

El-Husseini AE, Schnell E, Chetkovich DM, Nicoll RA, Bredt DS (2000) PSD-95 involvement in maturation of excitatory synapses. Science 290:1364–1368.

El-Husseini Ael D, Schnell E, Dakoji S, Sweeney N, Zhou Q, Prange O, Gauthier-Campbell C, Aguilera-Moreno A, Nicoll RA, Bredt DS (2002). Synaptic strength regulated by palmitate cycling on PSD-95. Cell 108:849–863.

Elgersma Y, Silva AJ (1999) Molecular mechanisms of synaptic plasticity and memory. Curr Opin Neurobiol 9:209–213.

Elias GM, Funke L, Stein V, Grant SG, Bredt DS, Nicoll RA (2006) Synapse-specific and developmentally regulated targeting of AMPA receptors by a family of MAGUK scaffolding proteins. Neuron 52:307–320.
Esteban JA (2004) Living with the enemy: a physiological role for the beta-amyloid peptide. Trends Neurosci 27:1–3.
Esteban JA, Shi SH, Wilson C, Nuriya M, Huganir RL. Malinow R (2003) PKA phosphorylation of AMPA receptor subunits controls synaptic trafficking underlying plasticity. Nature Neurosci 6:136–143.
Feldman DE, Nicoll RA, Malenka RC (1999) Synaptic plasticity at thalamocortical synapses in developing rat somatosensory cortex: LTP, LTD, and silent synapses. J Neurobiol 41:92–101.
Fischer M, Kaech S, Knutti D, Matus A (1998) Rapid actin-based plasticity in dendritic spines. Neuron 20:847–854.
Fleck MW (2006) Glutamate receptors and endoplasmic reticulum quality control: looking beneath the surface. Neuroscientist 12:232–244.
Fukata Y, Tzingounis AV, Trinidad JC, Fukata M, Burlingame AL, Nicoll RA, Bredt DS (2005) Molecular constituents of neuronal AMPA receptors. J Cell Biol 169:399–404.
Galdzicki Z, Siarey RJ (2003) Understanding mental retardation in Down's syndrome using trisomy 16 mouse models. Genes Brain Behav 2:167–178.
Gao C, Sun X, Wolf ME (2006) Activation of D1 dopamine receptors increases surface expression of AMPA receptors and facilitates their synaptic incorporation in cultured hippocampal neurons. J Neurochem 98:1664–1677.
Gerges NZ, Backos DS, Esteban JA (2004a) Local control of AMPA receptor trafficking at the postsynaptic terminal by a small GTPase of the Rab family. J Biol Chem 279:43870–43878.
Gerges NZ, Tran IC, Backos DS, Harrell JM, Chinkers M, Pratt WB, Esteban JA (2004b) Independent functions of hsp90 in neurotransmitter release and in the continuous synaptic cycling of AMPA receptors. J Neurosci 24:4758–4766.
Gerges NZ, Backos DS, Rupasinghe CN, Spaller MR, Esteban JA (2006) Dual role of the exocyst in AMPA receptor targeting and insertion into the postsynaptic membrane. Embo J 25:1623–1634.
Goel A, Jiang B, Xu LW, Song L, Kirkwood A, Lee HK (2006) Cross-modal regulation of synaptic AMPA receptors in primary sensory cortices by visual experience. Nature Neurosci 9:1001–1003.
Goldstein LS, Yang Z (2000) Microtubule-based transport systems in neurons: the roles of kinesins and dyneins. Annu Rev Neurosci 23:39–71.
Gomes AR, Cunha P, Nuriya M, Faro CJ, Huganir RL, Pires EV, Carvalho AL, Duarte CB (2004) Metabotropic glutamate and dopamine receptors co-regulate AMPA receptor activity through PKA in cultured chick retinal neurones: effect on GluR4 phosphorylation and surface expression. J Neurochem 90:673–682.
Gomes AR, Correia SS, Esteban JA, Duarte CB, Carvalho AL (2007) PKC anchoring to GluR4 AMPA receptor subunit modulates PKC-driven receptor phosphorylation and surface expression. Traffic 8:259–269.
Greger IH, Khatri L, Ziff EB (2002) RNA editing at arg607 controls AMPA receptor exit from the endoplasmic reticulum. Neuron 34:759–772.
Greger IH, Khatri L, Kong X, Ziff EB (2003) AMPA receptor tetramerization is mediated by Q/R editing. Neuron 40:763–774.
Greger IH, Ziff EB, Penn AC (2007) Molecular determinants of AMPA receptor subunit assembly. Trends Neurosci 30:407–416.
Groc L, Heine M, Cognet L, Brickley K, Stephenson FA, Lounis B, Choquet D (2004) Differential activity-dependent regulation of the lateral mobilities of AMPA and NMDA receptors. Nature Neurosci 7:695–696.
Guo W, Roth D, Walch-Solimena C, Novick P (1999) The exocyst is an effector for Sec4p, targeting secretory vesicles to sites of exocytosis. Embo J 18:1071–1080.

Hanley JG (2006) Molecular mechanisms for regulation of AMPAR trafficking by PICK1. Biochem Soc Trans 34:931–935.
Hanley JG, Henley JM (2005) PICK1 is a calcium-sensor for NMDA-induced AMPA receptor trafficking. Embo J 24:3266–3278.
Hanley JG, Khatri L, Hanson PI, Ziff EB (2002) NSF ATPase and alpha-/beta-SNAPs disassemble the AMPA receptor-PICK1 complex. Neuron 34:53–67.
Hattula K, Furuhjelm J, Tikkanen J, Tanhuanpaa K, Laakkonen P, Peranen J (2006) Characterization of the Rab8-specific membrane traffic route linked to protrusion formation. J Cell Sci 119:4866–4877.
Hayashi Y, Shi SH, Esteban JA, Piccini A, Poncer JC, Malinow R (2000) Driving AMPA receptors into synapses by LTP and CaMKII: requirement for GluR1 and PDZ domain interaction. Science 287:2262–2267.
Hebb DO (1949) Organization of behavior (New York, Wiley).
Hering H, Sheng M (2001) Dendritic spines: structure, dynamics and regulation. Nature Rev Neurosci 2:880–888.
Hollmann M, Heinemann S (1994) Cloned glutamate receptors. Annu Rev Neurosci 17:31–108.
Hoover KB, Bryant PJ (2000) The genetics of the protein 4.1 family: organizers of the membrane and cytoskeleton. Curr Opin Cell Biol 12:229–234.
Hsia AY, Malenka RC, Nicoll RA (1998) Development of excitatory circuitry in the hippocampus. J Neurophysiol 79:2013–2024.
Huber LA, Pimplikar S, Parton RG, Virta H, Zerial M, Simons K (1993) Rab8, a small GTPase involved in vesicular traffic between the TGN and the basolateral plasma membrane. J Cell Biol 123:35–45.
Isaac JT, Crair MC, Nicoll RA, Malenka RC (1997) Silent synapses during development of thalamocortical inputs. Neuron 18:269–280.
Ju W, Morishita W, Tsui J, Gaietta G, Deerinck TJ, Adams SR, Garner CC, Tsien RY, Ellisman MH, Malenka RC (2004) Activity-dependent regulation of dendritic synthesis and trafficking of AMPA receptors. Nature Neurosci 7:244–253.
Kato AS, Zhou W, Milstein AD, Knierman MD, Siuda ER, Dotzlaf JE, Yu H, Hale JE, Nisenbaum ES, Nicoll RA, Bredt DS (2007) New transmembrane AMPA receptor regulatory protein isoform, gamma-7, differentially regulates AMPA receptors. J Neurosci 27:4969–4977.
Katz LC, Shatz CJ (1996) Synaptic activity and the construction of cortical circuits. Science 274:1133–1138.
Kennedy MJ, Ehlers MD (2006) Organelles and trafficking machinery for postsynaptic plasticity. Annu Rev Neurosci 29:325–362.
Kim CH, Chung HJ, Lee HK, Huganir RL (2001) Interaction of the AMPA receptor subunit GluR2/3 with PDZ domains regulates hippocampal long-term depression. Proc Natl Acad Sci USA 98:11725–11730.
Kim CH, Lisman JE (1999) A role of actin filament in synaptic transmission and long-term potentiation. J Neurosci 19:4314–4324.
Ko J, Kim S, Valtschanoff JG, Shin H, Lee JR, Sheng M, Premont RT, Weinberg RJ, Kim E (2003) Interaction between liprin-alpha and GIT1 is required for AMPA receptor targeting. J Neurosci 23:1667–1677.
Kohler M, Kornau HC, Seeburg PH (1994) The organization of the gene for the functionally dominant alpha-amino-3-hydroxy-5-methylisoxazole-4-propionic acid receptor subunit GluR-B. J Biol Chem 269:17367–17370.
Kolleker A, Zhu JJ, Schupp BJ, Qin Y, Mack V, BorchardtT, Kohr G, Malinow R, Seeburg PH, Osten P (2003) Glutamatergic plasticity by synaptic delivery of GluR-B(long)-containing AMPA receptors. Neuron 40:1199–1212.
Konradi C, Heckers S (2003) Molecular aspects of glutamate dysregulation: implications for schizophrenia and its treatment. Pharmacol Ther 97:153–179.
Kopec CD, Li B, Wei W, Boehm J, Malinow R (2006) Glutamate receptor exocytosis and spine enlargement during chemically induced long-term potentiation. J Neurosci 26:2000–2009.

Kulangara K, Kropf M., Glauser L., Magnin S, Alberi S, Yersin A, Hirling H (2007). Phosphorylation of glutamate receptor interacting protein 1 regulates surface expression of glutamate receptors. J Biol Chem 282:2395–2404.

Kuusinen A, Abele R, Madden DR, Keinanen K (1999) Oligomerization and ligand-binding properties of the ectodomain of the alpha-amino-3-hydroxy-5-methyl-4-isoxazole propionic acid receptor subunit GluRD. J Biol Chem 274:28937–28943.

Lee SH, Liu L, Wang YT, Sheng M (2002) Clathrin adaptor AP2 and NSF interact with overlapping sites of GluR2 and play distinct roles in AMPA receptor trafficking and hippocampal LTD. Neuron 36:661–674.

Lee SH, Simonetta A, Sheng M (2004) Subunit rules governing the sorting of internalized AMPA receptors in hippocampal neurons. Neuron 43:221–236.

Letts VA, Felix R, Biddlecome GH, Arikkath J, Mahaffey CL, Valenzuela A, Bartlett FS 2nd, Mori Y, Campbell KP, Frankel WN (1998) The mouse stargazer gene encodes a neuronal Ca2+-channel gamma subunit. Nature Genet 19:340–347.

Leuschner WD, Hoch W (1999) Subtype-specific assembly of alpha-amino-3-hydroxy-5-methyl-4-isoxazole propionic acid receptor subunits is mediated by their n-terminal domains. J Biol Chem 274:16907–16916.

Ling DS, Benardo LS, Sacktor TC (2006) Protein kinase Mzeta enhances excitatory synaptic transmission by increasing the number of active postsynaptic AMPA receptors. Hippocampus 16:443–452.

Lise MF, Wong TP, Trinh A, Hines RM, Liu L, Kang R, Hines DJ, Lu J, Goldenring JR, Wang YT, El-Husseini A (2006) Involvement of myosin Vb in glutamate receptor trafficking. J Biol Chem 281:3669–3678.

Lisman JE, Zhabotinsky AM (2001) A model of synaptic memory: a CaMKII/PP1 switch that potentiates transmission by organizing an AMPA receptor anchoring assembly. Neuron 31:191–201.

Losi G, Prybylowski K, Fu Z, Luo JH, Vicini S (2002) Silent synapses in developing cerebellar granule neurons. J Neurophysiol 87:1263–1270.

Lu W, Ziff EB (2005) PICK1 interacts with ABP/GRIP to regulate AMPA receptor trafficking. Neuron 47:407–421.

Lu W, Man H, Ju W, Trimble WS, MacDonald JF, Wang YT (2001) Activation of synaptic NMDA receptors induces membrane insertion of new AMPA receptors and LTP in cultured hippocampal neurons. Neuron 29:243–254.

Luscher C, Xia H, Beattie EC, Carroll RC, von Zastrow M. Malenka RC, Nicoll RA (1999) Role of AMPA receptor cycling in synaptic transmission and plasticity. Neuron 24:649–658.

Malinow R, Malenka RC (2002). AMPA receptor trafficking and synaptic plasticity. Annu Rev Neurosci 25:103–126.

Malinow R, Mainen ZF, Hayashi Y (2000) LTP mechanisms: from silence to four-lane traffic. Curr Opin Neurobiol 10:352–357.

Man HY, Lin JW, Ju WH, Ahmadian G, Liu L, Becker LE, Sheng M, Wang YT (2000) Regulation of AMPA receptor-mediated synaptic transmission by clathrin-dependent receptor internalization. Neuron 25:649–662.

Man HY, Wang Q, Lu WY, Ju W, Ahmadian G, Liu L, D'Souza S, Wong TP, Taghibiglou C, Lu J, Becker LE, Pei L, Liu F, Wymann MP, MacDonald JF, Wang YT (2003). Activation of PI3-kinase is required for AMPA receptor insertion during LTP of mEPSCs in cultured hippocampal neurons. Neuron 38:611–624.

Man HY, Sekine-Aizawa Y, Huganir RL (2007) Regulation of {alpha}-amino-3-hydroxy-5-methyl-4-isoxazolepropionic acid receptor trafficking through PKA phosphorylation of the Glu receptor 1 subunit. Proc Natl Acad Sci USA 104:3579–3584.

Martin SJ, Grimwood PD, Morris RG (2000) Synaptic plasticity and memory: an evaluation of the hypothesis. Annu Rev Neurosci 23:649–711.

McGee AW, Bredt DS (2003). Assembly and plasticity of the glutamatergic postsynaptic specialization. Curr Opin Neurobiol 13:111–118.

Meng Y, Zhang Y, Jia Z (2003) Synaptic transmission and plasticity in the absence of AMPA glutamate receptor GluR2 and GluR3. Neuron 39:163–176.

Newey SE, Velamoor V, Govek EE, Van Aelst L (2005) Rho GTPases, dendritic structure, and mental retardation. J Neurobiol 64:58–74.

Nishimune A, Isaac JT, Molnar E, Noel J, Nash SR, Tagaya M, Collingridge GL, Nakanishi S, Henley JM (1998) NSF binding to GluR2 regulates synaptic transmission. Neuron 21:87–97.

Novick P, Medkova M, Dong G, Hutagalung A, Reinisch K, Grosshans B (2006) Interactions between Rabs, tethers, SNAREs and their regulators in exocytosis. Biochem Soc Trans 34:683–686.

Nowak L, Bregestovski P, Ascher P, Herbet A, Prochiantz A (1984) Magnesium gates glutamate-activated channels in mouse central neurones. Nature 307:462–465.

Osterweil E, Wells DG, Mooseker MS (2005) A role for myosin VI in postsynaptic structure and glutamate receptor endocytosis. J Cell Biol 168:329–338.

Park M, Penick EC, Edwards JG, Kauer JA, Ehlers MD (2004) Recycling endosomes supply AMPA receptors for LTP. Science 305:1972–1975.

Park M, Salgado JM, Ostroff L, Helton TD, Robinson CG, Harris KM, Ehlers MD (2006). Plasticity-induced growth of dendritic spines by exocytic trafficking from recycling endosomes. Neuron 52:817–830.

Passafaro M, Piech V, Sheng M (2001) Subunit-specific temporal and spatial patterns of AMPA receptor exocytosis in hippocampal neurons. Nature Neurosci 4:917–926.

Perez JL, Khatri L, Chang C, Srivastava S, Osten P, Ziff EB (2001) PICK1 targets activated protein kinase Calpha to AMPA receptor clusters in spines of hippocampal neurons and reduces surface levels of the AMPA-type glutamate receptor subunit 2. J Neurosci 21:5417–5428.

Petralia RS, Esteban JA, Wang YX, Partridge JG, Zhao HM, Wenthold RJ, Malinow R (1999) Selective acquisition of AMPA receptors over postnatal development suggests a molecular basis for silent synapses. Nature Neurosci 2:31–36.

Pickard L, Noel J, Duckworth JK, Fitzjohn SM, Henley JM, Collingridge GL, Molnar E (2001) Transient synaptic activation of NMDA receptors leads to the insertion of native AMPA receptors at hippocampal neuronal plasma membranes. Neuropharmacology 41:700–713.

Plant K, Pelkey KA, Bortolotto ZA, Morita D, Terashima A, McBain CJ, Collingridge GL, Isaac JT (2006) Transient incorporation of native GluR2-lacking AMPA receptors during hippocampal long-term potentiation. Nauret Neurosci 9:602–604.

Riefler GM, Balasingam G, Lucas KG, Wang S, Hsu SC, Firestein BL (2003) Exocyst complex subunit sec8 binds to postsynaptic density protein-95 (PSD-95): a novel interaction regulated by cypin (cytosolic PSD-95 interactor). Biochem J 373:49–55.

Rosenmund C, Stern-Bach Y, Stevens CF (1998) The tetrameric structure of a glutamate receptor channel. Science 280:1596–1599.

Rouach N, Byrd K, Petralia RS, Elias GM, Adesnik H, Tomita S, Karimzadegan S, Kealey C, Bredt DS, Nicoll RA (2005) TARP gamma-8 controls hippocampal AMPA receptor number, distribution and synaptic plasticity. Nature Neurosci 8:1525–1533.

Rowan MJ, Klyubin I, Cullen WK, Anwyl R (2003) Synaptic plasticity in animal models of early Alzheimer's disease. Philos Trans R Soc Lond B Biol Sci 358:821–828.

Rumpel S, LeDoux J, Zador A, Malinow R (2005) Postsynaptic receptor trafficking underlying a form of associative learning. Science 308:83–88.

Sans N, Prybylowski K, Petralia RS, Chang K, Wang YX, Racca C, Vicini S, Wenthold RJ (2003) NMDA receptor trafficking through an interaction between PDZ proteins and the exocyst complex. Nature Cell Biol 5:520–530.

Schluter OM, Xu W, Malenka RC (2006) Alternative N-terminal domains of PSD-95 and SAP97 govern activity-dependent regulation of synaptic AMPA receptor function. Neuron 51:99–111.

Schnell E, Sizemore M, Karimzadegan S, Chen L, Bredt DS, Nicoll RA (2002) Direct interactions between PSD-95 and stargazin control synaptic AMPA receptor number. Proc Natl Acad Sci USA 99:13902–13907.

Schulz TW, Nakagawa T, Licznerski P, Pawlak V, Kolleker A, Rozov A, Kim J, Dittgen T, Kohr G, Sheng M, Seeburg PH, Osten P. (2004) Actin/alpha-actinin-dependent transport of AMPA receptors in dendritic spines: role of the PDZ-LIM protein RIL. J Neurosci 24:8584–8594.

Seidenman KJ, Steinberg JP. Huganir R, Malinow R (2003) Glutamate receptor subunit 2 Serine 880 phosphorylation modulates synaptic transmission and mediates plasticity in CA1 pyramidal cells. J Neurosci 23:9220–9228.

Setou M, Seog DH, Tanaka Y, Kanai Y, Takei Y, Kawagishi M, Hirokawa N (2002) Glutamate-receptor-interacting protein GRIP1 directly steers kinesin to dendrites. Nature 417:83–87.

Shen L, Liang F, Walensky LD, Huganir RL (2000) Regulation of AMPA receptor GluR1 subunit surface expression by a 4. 1N-linked actin cytoskeletal association. J Neurosci 20:7932–7940.

Sheng M, Lee SH (2001) AMPA receptor trafficking and the control of synaptic transmission. Cell 105:825–828.

Shi S, Hayashi Y, Esteban JA, Malinow R (2001) Subunit-specific rules governing AMPA receptor trafficking to synapses in hippocampal pyramidal neurons. Cell 105:331–343.

Shin H, Wyszynski M, Huh KH, Valtschanoff JG, Lee JR, Ko J, Streuli M, Weinberg RJ, Sheng M, Kim E (2003) Association of the Kinesin Motor KIF1A with the Multimodular Protein Liprin-alpha. J Biol Chem 278:11393–11401.

Song I, Huganir RL (2002) Regulation of AMPA receptors during synaptic plasticity. Trends Neurosci 25:578–588.

Srivastava S, Osten P, Vilim FS, Khatri L, Inman G, States B, Daly C, DeSouza S, Abagyan R, Valtschanoff JG, Weinberg RJ, Ziff EB (1998) Novel anchorage of GluR2/3 to the postsynaptic density by the AMPA receptor-binding protein ABP. Neuron 21:581–591.

Steinberg JP, Takamiya K, Shen Y, Xia J, Rubio ME, Yu S, Jin W, Thomas GM, Linden DJ, Huganir RL (2006) Targeted in vivo mutations of the AMPA receptor subunit GluR2 and its interacting protein PICK1 eliminate cerebellar long-term depression. Neuron 49:845–860.

Steiner P, Alberi S, Kulangara K, Yersin A, Sarria JC, Regulier E, Kasas S, Dietler G, Muller D, Catsicas S, Hirling H (2005) Interactions between NEEP21, GRIP1 and GluR2 regulate sorting and recycling of the glutamate receptor subunit GluR2. Embo J 24:2873–2884.

Stephan KE, Baldeweg T, Friston KJ (2006) Synaptic plasticity and dysconnection in schizophrenia. Biol Psych 59:929–939.

Takahashi T, Svoboda K, Malinow R (2003) Experience strengthening transmission by driving AMPA receptors into synapses. Science 299:1585–1588.

Tardin C, Cognet L, Bats C, Lounis B, Choquet D (2003) Direct imaging of lateral movements of AMPA receptors inside synapses. Embo J 22:4656–4665.

Tichelaar W, Safferling M, Keinanen K, Stark H, Madden DR (2004) The three-dimensional structure of an ionotropic glutamate receptor reveals a dimer-of-dimers assembly. J Mol Biol 344:435–442.

Tomita S, Chen L, Kawasaki Y, Petralia RS, Wenthold RJ, Nicoll RA, Bredt DS (2003) Functional studies and distribution define a family of transmembrane AMPA receptor regulatory proteins. J Cell Biol 161:805–816.

Tomita S, Adesnik H, Sekiguchi M, Zhang W, Wada K, Howe JR, Nicoll RA, Bredt DS (2005) Stargazin modulates AMPA receptor gating and trafficking by distinct domains. Nature 435:1052–1058.

Turetsky D, Garringer E, Patneau DK (2005) Stargazin modulates native AMPA receptor functional properties by two distinct mechanisms. J Neurosci 25:7438–7448.

Turner PR, O'Connor K, Tate WP, Abraham WC (2003) Roles of amyloid precursor protein and its fragments in regulating neural activity, plasticity and memory. Prog Neurobiol 70:1–32.

Vandenberghe W, Nicoll RA, Bredt DS (2005) Interaction with the unfolded protein response reveals a role for stargazin in biosynthetic AMPA receptor transport. J Neurosci 25:1095–1102.

Wenthold RJ, Petralia RS, Blahos J II, Niedzielski AS (1996). Evidence for multiple AMPA receptor complexes in hippocampal CA1/CA2 neurons. J Neurosci 16:1982–1989.

Wu G, Malinow R, Cline HT (1996) Maturation of a central glutamatergic synapse. Science 274:972–976.
Wu H, Nash JE, Zamorano P, Garner CC (2002) Interaction of SAP97 with minus-end-directed actin motor myosin VI. Implications for AMPA receptor trafficking. J Biol Chem 277:30928–30934.
Wyszynski M, Kim E, Dunah AW, Passafaro M, Valtschanoff JG, Serra-Pages C, Streuli M, Weinberg RJ, Sheng M (2002) Interaction between GRIP and liprin-alpha/SYD2 is required for AMPA receptor targeting. Neuron 34:39–52.
Xia J, Zhang X, Staudinger J, Huganir RL (1999) Clustering of AMPA receptors by the synaptic PDZ domain-containing protein PICK1. Neuron 22:179–187.
Yasuda H, Barth AL, Stellwagen D, Malenka RC (2003) A developmental switch in the signaling cascades for LTP induction. Nature Neurosci 6:15–16.
Zerial M, McBride H (2001) Rab proteins as membrane organizers. Nature Rev Mol Cell Biol 2:107–117.
Zhu JJ, Esteban JA, Hayashi Y, Malinow R (2000) Postnatal synaptic potentiation: delivery of GluR4-containing AMPA receptors by spontaneous activity. Nature Neurosci 3:1098–1106.
Ziff EB (2007) TARPs and the AMPA receptor trafficking paradox. Neuron 53:627–633.
Ziv NE, Garner CC (2004) Cellular and molecular mechanisms of presynaptic assembly. Nature Rev Neurosci 5:385–399.

Imaging of Experience-Dependent Structural Plasticity in the Mouse Neocortex in vivo

Antony Holtmaat[1,3], *V. De Paola*[1,4], *L. Wilbrecht*[1,5], and *G. Knott*[2]

Summary. The functionality of adult neocortical circuits can be altered by novel experiences or learning. This functional plasticity appears to rely on changes in the strength of neuronal connections that were established during development. Here we will describe studies in which we have addressed whether structural changes, including the remodeling of axons and dendrites with synapse formation and elimination, could underlie experience-dependent plasticity in the adult neocortex. Using 2-photon laser-scanning microscopy transgenic mice expressing GFP in a subset of pyramidal cells, we have observed that a small subset of dendritic spines continuously appear and disappear on a daily basis, whereas the majority of spines persists for months. Axonal boutons from different neuronal classes displayed similar behavior, although the extent of remodeling varied. Under baseline conditions, new spines in the barrel cortex were mostly transient and rarely survived for more than a week. However, when every other whisker was trimmed (a paradigm known to induce adaptive functional changes in barrel cortex), the generation and loss of persistent spines was enhanced. Ultrastructural reconstruction of previously imaged spines and boutons using serial section electron microscopy showed that new spines slowly form synapses. New spines that persisted for a few days always had synapses, whereas very young spines often lacked synapses. New synapses were predominantly found on large, multisynapse boutons, suggesting that spine growth is followed by synapse formation, preferentially on existing boutons. Altogether our data indicate that novel sensory experience drives the stabilization of new spines on subclasses of cortical neurons and promotes the formation of new synapses. These synaptic changes likely underlie experience-dependent functional remodeling of specific neocortical circuits.

Introduction

During development, the formation of neural circuits in the mammalian neocortex is guided in a stereotypic way by a myriad of intracellular and extracellular molecular cues and is sculpted by spontaneous and sensory-evoked activity (Hua and Smith 2004; Katz and Shatz 1996). Although this so-called activity-dependent plasticity is most robust during development, neuronal circuits remain plastic in the adult brain. For example, cortical sensory maps can change in size and location upon peripheral lesions, including amputations, and changes in experience (Buonomano and Merzenich

[1] Cold Spring Harbor laboratory, Cold Spring Harbor, New York 11724, USA, email: holtmaat@cshl.edu

[2] Départment de Biologie Cellulaire et de Morphologie, University of Lausanne, 1005 Lausanne, Switzerland

[3] Current address: Départment des Neurosciences fondamentales, University of Geneva, 1211 Geneve, Switzerland

[4] Current address: MRC-Clinical Sciences Centre, Imperial College London, London W12 0NN, United Kingdom

[5] Current address: Keck Center, UCSF, San Francisco, California 94143, USA

Selkoe et al.
Synaptic Plasticity and the Mechanism of Alzheimer's Disease
© Springer-Verlag Berlin Heidelberg 2008

1998; Darian-Smith and Gilbert 1994; Daw et al. 1992; Diamond et al. 1994; Fox 2002; Frenkel et al. 2006; Gilbert 1998; Wang et al. 1995). Although this adult plasticity is thought to depend on changes in the strength of established synaptic connections (Hebb 1949), it could also involve structural alterations, including synapse formation and elimination (Antonini and Stryker 1993; Chklovskii et al. 2004; Knott et al. 2002; Lowel and Singer 1992; Ramon y Cajal 1893; Stepanyants et al. 2002; Turner and Greenough 1985; Ziv and Smith 1996). To address this possibility, we have focused our recent studies on dendritic spines and axonal boutons, the sites where the majority of excitatory synapses are found and, therefore, potential substrates for structural plasticity of cortical circuits.

Dendritic spines are tiny protrusions from the dendritic shaft, with volumes that can range from 0.001 to 1 μm^3 (Peters and Kaiserman-Abramof 1970; Sorra and Harris 2000). They can undergo fast structural remodeling on time scales of seconds and minutes (Bonhoeffer and Yuste 2002; Dailey and Smith 1996; Fischer et al. 1998; Lendvai et al. 2000; Matus 2000; Yuste and Bonhoeffer 2004), and de novo appearances and complete retractions can occur over hours and days (Dailey and Smith 1996; Engert and Bonhoeffer 1999; Holtmaat et al. 2005; Lendvai et al. 2000; Maletic-Savatic et al, 1999; Trachtenberg et al. 2002). They can emerge (Engert and Bonhoeffer 1999; Maletic-Savatic et al. 1999) or expand (Matsuzaki et al. 2004) in response to synaptic stimulation and make synapses (Knott et al. 2006; Knott et al. 2002; Toni et al. 1999, 2007). Dendritic spines could, therefore, be central to the brain's capacity to change its connectivity, increasing the dendrites' ability to connect with axons that are not in direct contact (Chklovskii et al. 2004; Peters and Kaiserman-Abramof 1970; Stepanyants and Chklovskii 2005; Swindale 1981).

Numerous studies that have manipulated sensory experience have shown changes in spine or synapse densities within the brain. These changes have been reported after rearing or extensive training in enriched environments (Beaulieu and Colonnier 1987; Moser 1999; Moser et al. 1994; Turner and Greenough 1985) but also after long-term sensory deprivation (Zuo et al. 2005b). Together with the observations that their dynamics are regulated by activity, these findings have led to the idea that dendritic spines are potential substrates, along with the axons, for activity-dependent plasticity in the adult brain (Bonhoeffer and Yuste 2002; Moser 1999; Segal 2005; Yuste and Bonhoeffer 2004).

Threaded amongst the dendrites and their spines are the sources of their inputs, the axons. Studded along the axonal branches are en passant boutons (EPBs), or bouton terminaux (TBs) at their endings. Like the spine, these structures have also shown the ability to remodel, both in vivo and in vitro. TBs have been proposed to subserve, in part, the formation of dendritic shaft synapses (McGuire et al. 1984), similar to the role of dendritic spines that form synapses with EPBs. Axonal endings and boutons can undergo profound structural changes during development (Antonini and Stryker 1993; Portera-Cailliau et al. 2005; Ruthazer et al. 2003). In the adult cortex, axonal sprouting has been observed after long-lasting peripheral lesions of sensory organs (Darian-Smith and Gilbert 1994; Florence et al. 1998). Many types of axonal boutons display rapid (minutes to hours) activity-dependent structural plasticity in dissociated (Colicos et al. 2001) and organotypic hippocampal cultures (De Paola et al. 2003; Nikonenko et al. 2003). The reorganization of axons at the micrometer level could play a role in circuit plasticity (Stepanyants et al. 2002). Similar to dendritic spines,

TBs and EPBs could also subserve structural plasticity, with their appearance and disappearance signaling synapse formation and elimination. In addition, changes in bouton size could reflect changes in synaptic weight (Murthy et al. 2001; Schikorski and Stevens 1997).

To investigate this possibility, we used in vivo two-photon laser scanning microscopy to monitor dendritic spines and axonal boutons in the mouse neocortex during changes in sensory experience (Figs. 1 and 2). Transgenic mice that express GFP or YFP in a subset of layer 5B and 2/3 cortical pyramidal cells were used, in which we imaged apical dendrites in the superficial cortical layers through a glass window

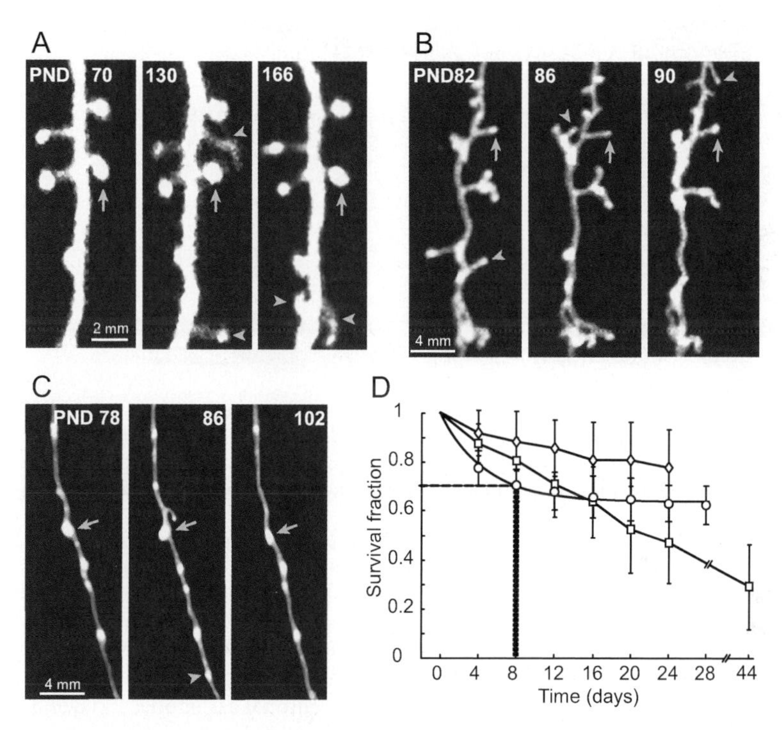

Fig. 1. Structural plasticity of dendritic spines and axonal boutons. **A.** Time-lapse image of a L5 pyramidal cell dendritic branch showing examples of transient (*arrowheads*) and persistent (*arrows*) spines. **B.** Time-lapse image of a presumed L6 pyramidal cell axon showing examples of appearing and disappearing TBs (*arrowheads*). *Arrows* point to a TB that remained stable over this time period. **C.** Time-lapse image of a presumed L2/3/5 pyramidal cell axon with EPBs. *Arrows* indicate stable boutons; the *arrowhead* indicates a bouton that appeared and subsequently disappeared. **D.** Survival functions of dendritic spines (*circles*), axonal L6 TBs (*squares*) and L2/3/5 EPBs (*diamonds*). The survival function of dendritic spines is fitted with a single exponential and a constant term.

Holtmaat AJ, Trachtenberg JT, Wilbrecht L, Shepherd GM, Zhang X, Knott GW, Svoboda K (2005) Transient and persistent dendritic spines in the neocortex in vivo. Neuron 45:279–291

De Paola V, Holtmaat A, Knott G, Song S, Wilbrecht L, Welker E, Caroni P, Svoboda K (2006) Cell type-specific structural plasticity of axonal branches and boutons in the adult neocortex. Neuron 49:861–875

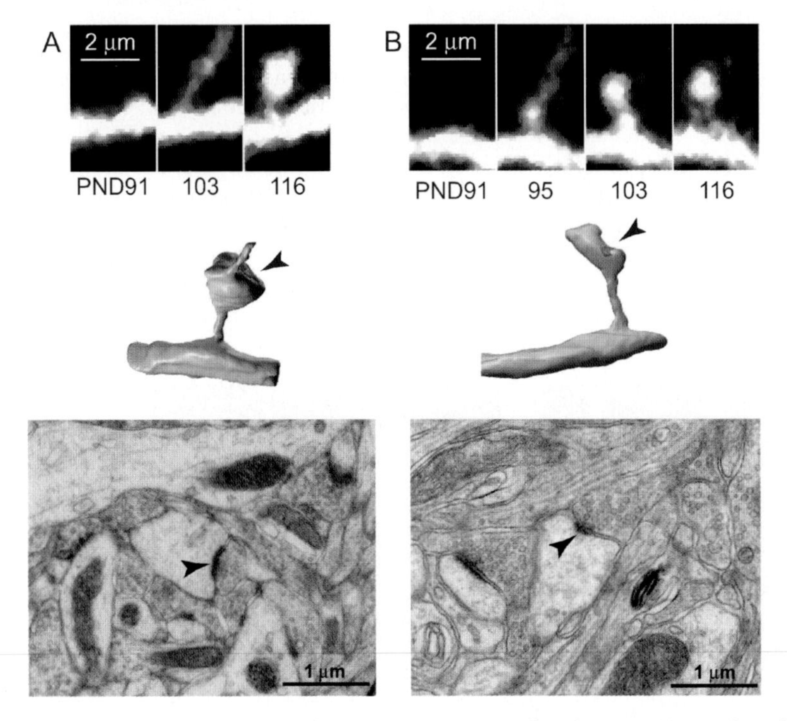

Fig. 2. New persistent spines make synapses. **A, B.** Examples of new persistent spines after whisker trimming. *Upper panels* show the time-lapse images, the *middle panels* the 3D-reconstructions from serial EM sections, and the *lower panels* the single EM sections. *Arrowheads* indicate asymmetric synapses. Spine A is between 14 and 18 days old; spine B is between 22 and 26 days old.
Knott GW, Holtmaat A, Wilbrecht L, Welker E, Svoboda K (2006) Spine growth precedes synapse formation in the adult neocortex in vivo. Nat Neurosci 9:1117–1123

glued in the skull. By placing the window over the primary somatosensory cortex, we were able to focus on neurons situated within the whisker sensory map, or barrel cortex, characterized by the unique arrangement of multineuronal assemblies in layer 4, each receiving information from a single whisker at the periphery. Changes in sensory activity at the periphery can be elicited by the simple manipulation of whiskers, e.g., trimming, leading to perturbations in the response properties of cells (Diamond et al. 1993; Feldman and Brecht 2005; Fox and Wong 2005). We used this paradigm to investigate to what extent this action would cause changes in the structure and connectivity of neurons in this region.

Plasticity of Dendritic Spines

Our first experiments studied dendrites and their spines towards the end of the first postnatal month. Imaging dendrites from P16 through P28, we found that thin protrusions appeared and disappeared on a daily basis (Holtmaat et al. 2005). Most transient protrusions were long and thin, similar to filopodia described in earlier studies (Dailey and Smith 1996; Lendvai et al. 2000; Portera-Cailliau et al. 2003), whereas thick spines with a clear head tended to be stable. A quantitative analysis of the spine turnover showed that it was imbalanced, resulting in a net pruning of spines. A similar decline in spine density was observed in fixed tissue at these ages (Holtmaat et al. 2005) and in an independent in vivo imaging study (Zuo et al. 2005a). These results confirmed the notion that synapse densities in cortical layer 1 peak around birth or during postnatal development and gradually decline towards adulthood (De Felipe et al. 1997; Markus and Petit 1987; Rakic et al. 1986). However, the timing of excessive spine pruning that we observed did not exactly match that of the loss of synapses in barrel cortex as observed by electron microscopy (De Felipe et al. 1997). This discrepancy could be explained by the fact that earlier studies did not differentiate between subpopulations of pyramidal cells that were potentially in different developmental stages, as well as by the possibility that spines and synapses are not necessarily present in a one-to-one fashion at this age.

These results lead us to investigate to what extent spines continue to be generated and eliminated after development and whether the architecture of spiny dendrites is stable in adulthood. We imaged spines daily to asses their dynamics in both young adult (2–3 months old) and mature adult (6 months old) mice (Holtmaat et al. 2005; Trachtenberg et al. 2002). We found that although spines were still being generated and eliminated (Fig. 1A), the dynamics were much slower than during development. On average about 15% of the protrusions appeared or disappeared from day to day in mature adult mice (Fig. 1D). The largest fraction (>70%) of spines was stable over very long time periods (up to 3 months of imaging). The majority of the new spines was only present transiently and was usually eliminated within a few days. Spines that were seen for more than 4 days were very likely to be present for much longer times. We therefore concluded that spines in the adult cortex can be subdivided into transient spines and persistent spines. We defined spines with a lifespan of less than four days as transient spines and those with a lifespan of more than eight days as persistent spines (Holtmaat et al. 2005, 2006).

Characterizing the structure of transient and persistent spines from the fluorescence images showed that transient spines tended to be thin and long, whereas persistent spines were often large (Holtmaat et al. 2005). Transient spines were typically similar to the dendritic filopodia seen in young, growing dendrites (Dailey and Smith 1996; Lendvai et al. 2000; Portera-Cailliau et al. 2003) and bore similarities to the thin spines described in the hippocampus (Harris et al. 1992; Sorra and Harris 2000). We found similar spine morphologies in naïve perfusion-fixed brains, and their distributions were indistinguishable from in vivo imaged tissue, indicating that small-volume spines were not induced by imaging. Since spine volumes are proportional to the area of the postsynaptic density, AMPA receptor content, and the size of the presynapse, spine size is likely to correlate with synaptic strength (Kharazia and Weinberg 1999; Nusser et al. 1998; Takumi et al. 1999). Therefore, excitation of pyramidal cells is probably driven mainly through the synapses on persistent spines, with the new spines contributing little to the overall excitability and possibly even associated with postsynaptically silent synapses (Malinow and Malenka 2002).

Structural Plasticity of Axons

In another, but similar, imaging study, we characterized and followed different types of cortical axons over time (Fig. 1B, C; De Paola et al. 2006). In Thy-1-GFP (line M) transgenic mice, a variety of morphologically distinct axons could be encountered in layers 1 and 2 of the somatosensory cortex. The axons often had different branching profiles, an indication that they were derived from different neuronal subtypes. Characterizing morphological features and tracing the axons back to their origin, or to the white matter, showed that at least three different types of axons were present in the upper cortical layers: 1) axons that are presumably originating from the thalamus, with thick axonal shafts, frequent short branches, prominent varicosities at their tips, and occasional extremely large EPBs; 2) layer 6 pyramidal cell axons with thin shafts and a high density of thin, spine-like TBs; and 3) axons originating from pyramidal cells in layer 5 and 2/3 with thin shafts and a high density of small EPBs.

The large-scale structure of axonal arbors of all three of these classes was stable for long periods, with a 4% incidence of branch elimination or addition over one month. However, these findings concerned very short branches with an average length of ~10 μm. The fractional length change of entire axonal arbors in L1 was only on the order of 5% over 24 days, showing that the total axonal length is relatively stable. The axonal branches that grew and retracted often did so over the same trajectories. The movements were usually on the order of several micrometers every four days. However, the average branch length was constant over time, implying that elongation of axons was compensated for by retraction at other locations. The highest motility of axonal branches was encountered at the most distal endings of thalamocortical and L6 axons (on some occasions, up to 10 to 30 μm over four days). Distal endings retracted and elongated over nonrepeating and tortuous paths, presumably as the axon explored the neighborhood.

Putative synaptic boutons, both EPBs and TBs, also displayed dynamic behavior (De Paola et al. 2006). An independent in vivo imaging study in adult monkey visual cortex corroborated these results (Stettler et al. 2006). Although the densities of boutons were constant, some boutons appeared or disappeared over time periods of days. Scoring their presence and analyzing their lifetimes in a similar way to spines, we saw that the TBs (present on little stalks) behaved in a very similar way to spines: de novo appearing and completely retracting (Fig. 1B). The presence of EPBs was scored based on their brightness. Swellings were considered to be boutons if their brightness exceeded the axonal backbone with a factor 3. Retrospective electron microscopy confirmed that these types of boutons have synapses (De Paola et al. 2006). Losses were scored if the brightness dropped below 1.3 times the axonal backbone brightness (Fig. 1C). The vast majority of axonal boutons, both TBs and EPBs on thalamocortical axons, was stable over ~1 month of imaging (~85%). EPBs on intracortical axons displayed slightly higher dynamic behavior (Fig. 1C, D; ~77% stability). TBs on L6 axons were even more dynamic. The survival function had a linear shape, dropping to ~28% after 1.5 months of imaging (Fig. 1D), showing that the entire population of L6 TBs could turn over several times within the lifetime of the animal. The relatively greater dynamics of L6 TBs when compared to the relative stability of thalamocortical or L2/3 and L5 pyramidal cell EPBs indicates that structural and thereby functional plasticity likely varies between different cortical circuits.

Experience-Dependent Generation of Persistent Spines

The topographic map of the body in the somatosensory cortex can be perturbed both structurally and functionally by disruption of sensory inputs or new sensory experiences, which induce adaptive changes in the organization of the functional map; (Buonomano and Merzenich 1998; Gilbert 1998). For example, the loss of a subset of whiskers will result in a reduction of the cortical area in which the lost whiskers provide the dominant sensory input and, subsequently, an increase in the number of neurons that strongly respond to deflection of the spared whiskers (Diamond et al. 1993; Feldman and Brecht 2005; Fox and Wong 2005). Given that whisker trimming has not been reported to affect neurogenesis or neuronal survival in the adult barrel cortex, changes in receptive fields are likely to be the result of synaptic alterations in cortical circuitry (Fox 2002; Fox and Wong 2005). Growth and retractions of dendritic spines with synapse formation and elimination could be a mechanism to rebalance whisker representations in the barrel cortex. We tested whether the generation and loss of spines are enhanced by alternated whisker trimming (Holtmaat et al. 2006). Spines were tracked over a month by imaging every 4 days, and we focused our analysis on persistent spines (present for eight days or more). We categorized persistent spines as "always present" (AP; present for the entire imaging period), new persistent (NP; appearing during imaging and remaining for the for the last eight days of imaging) or lost persistent (LP; present for the first eight days of imaging and subsequently disappearing). Under baseline conditions, the fractions of NP (NP/(NP+AP)) and LP (LP/(LP+AP)) spines over a 28-day period were very small, on the order of 5 to 10%. However, when whiskers were trimmed after the third imaging time point (day 8), the fractions of both NP and LP spines increased to $\sim$15%. The NP spines were probably recruited from the transient spine population, since they often appeared as small and thin spines at first, and then gained volume over time (Fig. 2A, B). The fraction of NP spines was highly variable from neuron to neuron. Morphological analysis of the pyramidal cell dendrites showed that NP spines were preferentially added on apical dendritic tufts with a complex morphology, i.e., with many branches (more than seven branch orders with an average total length of $\sim$4.25 mm) and a first bifurcation at an average of 315 μm below the pial surface. These neurons resembled previously described intrinsically bursting L5B neurons (IB; Chagnac-Amitai et al. 1990; Larkman and Mason 1990; Tsiola et al. 2003). In contrast, apical dendrites with a very simple morphological appearance (fewer than seven branch orders with a total average length of $\sim$1.5 mm and a bifurcation within 150 μm below the pial surface) displayed a relatively low number of spine gains. These neurons resembled regular spiking neurons (RS; Chagnac-Amitai et al. 1990; Larkman and Mason 1990; Tsiola et al. 2003). The LP spines were equally distributed among complex- and simple-tuft cells. Therefore, simple-tuft dendrites displayed a net loss of persistent spines whereas complex-tuft dendrites displayed a net gain of persistent spines. This finding aligns with physiological measurements showing that RS (simple-tuft) cells tend to lose intracortical input, whereas IB (complex tuft) cells tend to gain cortical input upon trimming of a subset of whiskers (Petreanu et al. 2005, Wright and Fox, 2006).

New Persistent Spines form Synapses

Dendritic spines appear in many different forms, ranging from long and thin to stubby and large mushroom-like structures (Sorra and Harris 2000). In addition, not all spines have synapses (Arellano et al. 2007; Spacek 1982; Trachtenberg et al. 2002) and some spines have more than one synapse, which can be both excitatory or inhibitory (or a combination of the two) (Knott et al. 2002). It is not well understood to what extent these structurally different spines relate to different degrees of maturation and stages of synaptogenesis. To investigate the relationship between new spines, their maturation and synaptogenesis, we performed an ultrastructural analysis of persistent and new spines of various ages that were detected in vivo, including their surrounding neuropil (Knott et al. 2006). A random subset of previously imaged dendrites and their spines was identified in fixed tissue and completely reconstructed in three dimensions using serial-section electron microscopy (ssEM; Fig. 2). All spines that were seen on the last imaging day could be located along the reconstructed dendrite. We also reconstructed many axonal boutons in the proximity of dendritic spines of interest. Ultrastructural features were marked, such as the presence of post-synaptic densities (PSDs) on spines and dendritic shafts, presynaptic vesicles, mitochondria, smooth endoplasmic reticula (SER, and spine aparati) and supporting glial processes.

Spines were considered to have synapses if they had a PSD and an active zone (indicated by docked pre-synaptic vesicles) directly apposed to the PSD. Of all spines that were seen for the first time on the last imaging day – and were therefore likely to be at most a couple of days old – only ~30% bore a synapse. All new spines that were more than four days old had synapses, similar to the spines that were seen during the entire imaging period of one month (Fig. 2). We also analyzed the structural maturation of new spines by comparing the size and ultrastructure of spines of different ages. Young spines without synapses had smaller volumes and larger surface-to-volume ratios than new spines with synapses. The PSD area of the new synaptic spines was highly correlated with their volumes. Taken together, the data suggest that spines form as relatively thin protrusions and continue to mature by adding synaptic components and increasing their volume. Since larger PSDs are likely to contain more glutamate receptors (Kharazia and Weinberg 1999; Nusser et al. 1998; Takumi et al. 1999), this spine growth and maturation might reflect increasing synaptic strength. Some new spines also added SER or a spine apparatus, indicating that maturation is accompanied by an increased occurrence of SER, which might influence the spine's capacity to buffer calcium and regulate plasticity (Deller et al. 2007).

Most of the new spines that formed synapses made synaptic contacts with large axonal boutons that had one or more other synapses, the so called multi-synapse boutons (MSBs). Only about one third of the new spines made synapses with single synapse boutons (SSBs). SSBs could have been formed de novo, shortly before or upon contact with the new spine. They could have also arisen from pre-existing MSBs after one or more of the other synapses was pruned. Similarly, MSBs could have been formed de novo by a simultaneous contact of multiple new spines or could be the result of a pre-existing SSB on which one synapse or more was added. If we assume, however, that most of our samples represented the situation immediately or shortly after synaptogenesis, we concluded that most new spines (~65%) prefer to make synapses on existing boutons, whereas only ~35% make synapses with new boutons. These numbers are consistent with the turnover ratios of axonal boutons, which are typically lower than those of dendritic spines (De Paola et al. 2006; Holtmaat et al. 2005)

Discussion

The plasticity of dendritic spines and axonal boutons has been postulated to be a substrate for experience-dependent changes in cortical circuitry. In vitro studies in brain slices show that the morphology of spines and boutons is dynamic. With the use of long-term 2PLSM in vivo, recent studies have shown that the structural plasticity of dendritic spines and axonal boutons, including synapse formation and elimination, is an ongoing process that continues after development until far into adulthood (De Paola et al. 2006; Holtmaat et al. 2005, 2006). These findings suggest that, although the degree of structural plasticity decreases after postnatal development, activity-dependent mechanisms continue to sculpt synaptic connectivity in the adult neocortex.

We have found that a subset of spines and boutons appears and disappears in the somatosensory mouse cortex. The majority of spines are stable and, of the spines that are new, only a very small percentage, on the order of 5%, will be added to the persistent spine population under baseline conditions. We favor the model that, under naïve conditions, dendrites continuously attempt to make new contacts with axons, or vice versa (Fig. 3). However, since the connectivity within cortical circuits has presumably been optimized during development, the majority of these attempts fail to produce long-lasting contacts, perhaps due to homeostatic forces that limit the synaptic weight between established neuronal pairs and to a lack of correlated firing of neurons that are not already connected. This constant sampling would allow a continuous optimization of connectivity and prepare the circuit for prompt adaptations to changes in sensory input. The working hypothesis incorporates the idea that whenever the input and output of a particular circuit become unbalanced – for example, in the barrel cortex due to loss of subsets of whiskers – some new contacts might be favored over old, now dysfunctional contacts, resulting in a loss of a subset of old spines and a gain of NP spines (Fig. 3). We have found evidence in vivo of the existence of such a sample-and-selection mechanism in the adult barrel cortex (Holtmaat et al. 2006; Knott et al. 2006).

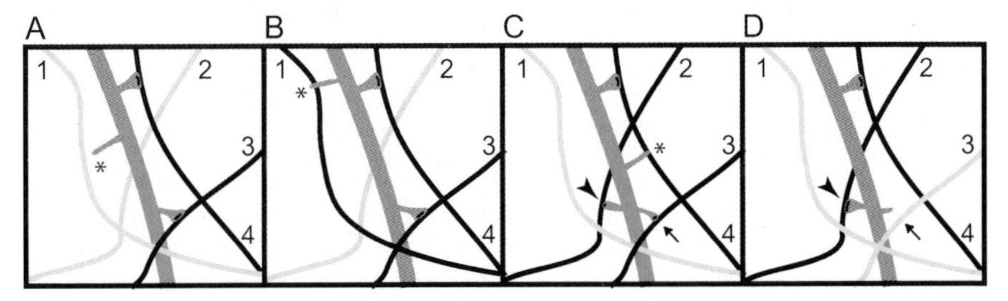

Fig. 3. Schematic of the sample and selection model. **A–D.** Schematic time-lapse image of a dendrite with spines (*gray*) and several surrounding axons that are either connected (*black*) by spines or unconnected (*light gray*). Synapses are indicated by *small black lines* within the spines. Transient spines are indicated by *asterisks*. **C, D.** Gains (*arrowheads*) and losses (*arrows*) of dendritic spines could result in long-term changes in the connectivity of neurons

Others have also observed experience-dependent changes in spine survival and pruning during postnatal development (Zuo et al. 2005b). Further experiments are necessary to determine which factors are critical for the generation of new spines and synapses, whether these factors are intrinsic or derived from presynaptic sites, whether their role is activity dependent, and whether synaptogenesis relates to LTP-like phenoma, spike timing-dependent mechanisms or other forms of correlated firing. A gain in the understanding of the mechanisms that underlie structural synaptic plasticity in the adult brain could help in the design of strategies to treat neurodegenerative diseases affecting synapses, such as Alzheimer's disease and stroke.

References

Antonini A, Stryker MP (1993) Rapid remodeling of axonal arbors in the visual cortex. Science 260:1819–1821.

Arellano JI, Espinosa A, Fairen A, Yuste R, De Felipe J (2007) Non-synaptic dendritic spines in neocortex. Neuroscience 145:464–469.

Beaulieu C, Colonnier M (1987) Effect of the richness of the environment on the cat visual cortex. J Comp Neurol 266:478–494.

Bonhoeffer T, Yuste R (2002) Spine motility. Phenomenology, mechanisms, and function. Neuron 35:1019–1027.

Buonomano DV, Merzenich MM (1998) Cortical plasticity: from synapses to maps. Annu Rev Neurosci 21:149–186.

Chagnac-Amitai Y, Luhmann HJ, Prince DA (1990) Burst generating and regular spiking layer 5 pyramidal neurons of rat neocortex have different morphological features. J Comp Neurol 296:598–613.

Chklovskii DB, Mel BW, Svoboda K (2004) Cortical rewiring and information storage. Nature 431:782–788.

Colicos MA, Collins BE, Sailor MJ, Goda Y (2001) Remodeling of synaptic actin induced by photoconductive stimulation. Cell 107:605–616.

Dailey ME, Smith SJ (1996) The dynamics of dendritic structure in developing hippocampal slices. J Neurosci 16:2983–2994.

Darian-Smith C, Gilbert CD (1994) Axonal sprouting accompanies functional reorganization in adult cat striate cortex. Nature 368:737–740.

Daw NW, Fox K, Sato H, Czepita D (1992) Critical period for monocular deprivation in the cat visual cortex. J Neurophysiol 67:197–202.

De Felipe J, Marco P, Fairen A, Jones EG (1997). Inhibitory synaptogenesis in mouse somatosensory cortex. Cereb Cortex 7:619–634.

De Paola V, Arber S, Caroni P (2003) AMPA receptors regulate dynamic equilibrium of presynaptic terminals in mature hippocampal networks. Nature Neurosci 6:491–500.

De Paola V, Holtmaat A, Knott G, Song S, Wilbrecht L, Caroni P, Svoboda K (2006) Cell type-specific structural plasticity of axonal branches and boutons in the adult neocortex. Neuron 49:861–875.

Deller T, Orth CB, Del Turco D, Vlachos A, Burbach GJ, Drakew A, Chabanis S, Korte M, Schwegler H, Haas CA, Frotscher M (2007) A role for synaptopodin and the spine apparatus in hippocampal synaptic plasticity. Ann Anat 189:5–16.

Diamond ME, Armstrong-James M, Ebner FF (1993) Experience-dependent plasticity in adult barrel cortex. Proc Natl Acad Sci USA 90:2082–2086.

Diamond ME, Huang W, Ebner FF (1994) Laminar comparison of somatosensory cortical plasticity. Science 265:1885–1888.

Engert F, Bonhoeffer T (1999) Dendritic spine changes associated with hippocampal long-term synaptic plasticity. Nature 399:66–70.

Feldman DE, Brecht M (2005) Map plasticity in somatosensory cortex. Science 310:810–815.
Fischer M, Kaech S, Knutti D, Matus A (1998) Rapid actin-based plasticity in dendritic spines. Neuron 20:847–854.
Florence SL, Taub HB, Kaas JH (1998) Large-scale sprouting of cortical connections after peripheral injury in adult macaque monkeys. Science 282: 1117–1121.
Fox K (2002) Anatomical pathways and molecular mechanisms for plasticity in the barrel cortex. Neuroscience 111:799–814.
Fox K, Wong RO (2005) A comparison of experience-dependent plasticity in the visual and somatosensory systems. Neuron 48:465–477.
Frenkel MY, Sawtell NB, Diogo AC, Yoon B, Neve RL, Bear MF (2006) Instructive effect of visual experience in mouse visual cortex. Neuron 51:339–349.
Gilbert CD (1998) Adult cortical dynamics. Physiol Rev 78:467–485.
Harris KM, Jensen FE, Tsao B (1992) Three-dimensional structure of dendritic spines and synapses in rat hippocampus (CA1) at postnatal day 15 and adult ages: implications for the maturation of synaptic physiology and long-term potentiation. J Neurosci 12:2685–2705.
Hebb DO (1949) The organization of behavior. New York: Wiley.
Holtmaat A, Wilbrecht L, Knott GW, Welker E, Svoboda K (2006) Experience-dependent and cell-type-specific spine growth in the neocortex. Nature 441:979–983.
Holtmaat A, Trachtenberg JT, Wilbrecht L, Shepherd GM, Zhang X, Knott GW, Svoboda K (2005) Transient and persistent dendritic spines in the neocortex in vivo. Neuron 45:279–291.
Hua JY, Smith SJ (2004) Neural activity and the dynamics of central nervous system development. Nature Neurosci 7:327–332.
Katz LC, Shatz CJ (1996) Synaptic activity and the construction of cortical circuits. Science 274:1133–1138.
Kharazia VN, Weinberg RJ (1999) Immunogold localization of AMPA and NMDA receptors in somatic sensory cortex of albino rat. J Comp Neurol 412:292–302.
Knott GW, Quairiaux C, Genoud C, Welker E (2002) Formation of dendritic spines with GABAergic synapses induced by whisker stimulation in adult mice. Neuron 34:265–273.
Knott GW, Holtmaat A, Wilbrecht L,Welker E, Svoboda K (2006) Spine growth precedes synapse formation in the adult neocortex in vivo. Nat Neurosci. 9:1117–1123.
Larkman A, Mason A (1990) Correlations between morphology and electrophysiology of pyramidal neurons in slices of rat visual cortex: I. Establishment of cell classes. J Neurosci. 10:1415–1428.
Lendvai B, Stern E, Chen B, Svoboda K (2000) Experience-dependent plasticity of dendritic spines in the developing rat barrel cortex in vivo. Nature 404: 876–881.
Lowel S, Singer W (1992) Selection of intrinsic horizontal connections in the visual cortex by correlated neuronal activity. Science 255:209–212.
Maletic-Savatic M, Malinow R, Svoboda K (1999) Rapid dendritic morphogenesis in CA1 hippocampal dendrites induced by synaptic activity. Science 283:1923–1927.
Malinow R, Malenka RC (2002) AMPA receptor trafficking and synaptic plasticity. Annu Rev Neurosci 25:103–126.
Markus EJ, Petit TL (1987) Neocortical synaptogenesis, aging, and behavior: lifespan development in the motor-sensory system of the rat. Exp Neurol 96:262–278.
Matsuzaki M, Honkura N, Ellis-Davies GC, Kasai H (2004) Structural basis of long-term potentiation in single dendritic spines. Nature 429:761–766.
Matus A (2000) Actin-based plasticity in dendritic spines. Science 290: 754–758.
McGuire BA, Hornung JP, Gilbert CD, Wiesel TN (1984) Patterns of synaptic input to layer 4 of cat striate cortex. J Neurosci 4:3021–3033.
Moser MB (1999) Making more synapses: a way to store information? Cell Mol Life Sci 55:593–600.
Moser MB, Trommwald M, Andersen P (1994) An increase in dendritic spine density on hippocampal CA1 cells following spatial-learning in adult rats suggests the formation of new synapses. Proc Natl Acad Sci USA 91:12673–12675.

Murthy VN, Schikorski T, Stevens CF, Zhu Y (2001) Inactivity produces increases in neurotransmitter release and synapse size. Neuron 32:673–682.
Nikonenko I, Jourdain P, Muller D (2003) Presynaptic remodeling contributes to activity-dependent synaptogenesis. J Neurosci 23:8498–8505.
Nusser Z, Lujan R, Laube G, Roberts JD, Molnar E, Somogyi P (1998) Cell type and pathway dependence of synaptic AMPA receptor number and variability in the hippocampus. Neuron 21:545–559.
Peters A, Kaiserman-Abramof IR (1970) The small pyramidal neuron of the rat cerebral cortex. The perikaryon, dendrites and spines. Am J Anat 127:321–355.
Petreanu LT, Shepherd GMG, Svoboda K (2005) Laser-scanning photostimulation reveals that two classes of layer 5B neurons mediate distinct aspects of experience-dependent plasticity. Soc Neurosci 35nd Annual Meeting 985.2.
Portera-Cailliau C, Pan DT, Yuste R (2003) Activity-regulated dynamic behavior of early dendritic protrusions: evidence for different types of dendritic filopodia. J Neurosci 23:7129–7142.
Portera-Cailliau C, Weimer RM, Paola VD, Caroni P, Svoboda K (2005) Diverse modes of axon elaboration in the developing neocortex. PLoS Biol 3:e272.
Rakic P, Bourgeois JP, Eckenhoff MF, Zecevic N, Goldman-Rakic PS (1986) Concurrent overproduction of synapses in diverse regions of the primate cerebral cortex. Science 232:232–235.
Ramon y Cajal S (1893) Neue Darstellung vom histologischen Bau des Centralnervensystems. Arch Anat Physiol Anat Abt Suppl:319–428.
Ruthazer , Akerman CJ, Cline HT (2003) Control of axon branch dynamics by correlated activity in vivo. Science 301:66–70.
Schikorski T, Stevens CF (1997) Quantitative ultrastructural analysis of hippocampal excitatory synapses. J Neurosci 17:5858–5867.
Segal M (2005) Dendritic spines and long-term plasticity. Nature Rev Neurosci 6:277–284.
Sorra KE, Harris KM (2000) Overview on the structure, composition, function, development, and plasticity of hippocampal dendritic spines. Hippocampus 10:501–511.
Spacek J (1982) 'Free' postsynaptic-like densities in normal adult brain: their occurrence, distribution, structure and association with subsurface cisterns. J Neurocytol 11:693–706.
Stepanyants A, Chklovskii DG (2005) Neurogeometry and potential synaptic connectivity. Trends Neurosci 28:387–394.
Stepanyants A, Hof PR, Chklovskii DB (2002) Geometry and structural plasticity of synaptic connectivity. Neuron 34:275–288.
Stettler DD, Yamahachi H, Li W, Denk W, Gilbert CD (2006) Axons and synaptic boutons are highly dynamic in adult visual cortex. Neuron 49:877–887.
Swindale NV (1981) Dendritic spines only connect. Trends Neurosci 4:240–241.
Takumi Y, Ramirez-Leon V, Laake P, Rinvik E, Ottersen OP (1999) Different modes of expression of AMPA and NMDA receptors in hippocampal synapses. Nature Neurosci 2:618–624.
Toni N, Buchs PA, Nikonenko I, Bron CR, Muller D (1999) LTP promotes formation of multiple spine synapses between a single axon terminal and a dendrite. Nature 402:421–425.
Toni N, Teng EM, Bushong EA, Aimone JB, Zhao C, Consiglio A, van Praag H, Martone ME, Ellisman MH, Gage FH (2007) Synapse formation on neurons born in the adult hippocampus. Nature Neurosci 10:727–734.
Trachtenberg JT, Chen BE, Knott GW, Feng G, Sanes JR, Welker E, Svoboda K (2002) Long-term in vivo imaging of experience-dependent synaptic plasticity in adult cortex. Nature 420:788–794.
Tsiola A, Hamzei-Sichani F, Peterlin Z, Yuste R (2003) Quantitative morphologic classification of layer 5 neurons from mouse primary visual cortex. J Comp Neurol 461:415–428.
Turner AM, Greenough WT (1985) Differential rearing effects on rat visual cortex synapses. I. Synaptic and neuronal density and synapses per neuron. Brain Res 329:195–203.
Wang X, Merzenich MM, Sameshima K, Jenkins WM (1995) Remodelling of hand representation in adult cortex determined by timing of tactile stimulation. Nature 378:71–75.

Wright NF, Fox K (2006) The time course of experience-depression in layer V of the rodent barrel cortex in vivo. Soc Neurosci 36nd Annual Meeting 53.27.
Yuste R, Bonhoeffer, T (2004) Genesis of dendritic spines: insights from ultrastructural and imaging studies. Nature Rev Neurosci 5:24–34.
Ziv NE, Smith SJ (1996) Evidence for a role of dendritic filopodia in synaptogenesis and spine formation. Neuron 17:91–102.
Zuo Y, Lin A, Chang P, Gan WB (2005a) Development of long-term dendritic spine stability in diverse regions of cerebral cortex. Neuron 46:181–189.
Zuo Y, Yang G, Kwon E, Gan WB (2005b) Long-term sensory deprivation prevents dendritic spine loss in primary somatosensory cortex. Nature 436:261–265.

Synapse Loss, Synaptic Plasticity and the Postsynaptic Density

Morgan Sheng[1]

Summary. Excitatory synapses of the mammalian brain usually occur on dendritic spines, the postsynaptic compartment of most excitatory synapses. The postsynaptic membrane contains a high concentration of NMDA receptors (NMDARs) and AMPA receptors (AMPARs) as well as their associated signaling proteins, which are assembled by scaffold proteins into the postsynaptic density (PSD). Composed of hundreds of distinct proteins, the PSD dynamically changes its structure and composition in response to synaptic activity. A comprehensive, quantitative and three-dimensional view of PSD architecture is gradually emerging, providing unprecedented details on the protein composition and stoichiometry of the PSD. Such knowledge facilitates understanding of the postsynaptic signaling mechanisms that control the strengthening and growth versus the weakening and loss of synapses. The molecular organization of the PSD reveals several signaling pathways that likely mediate synaptic depression and synapse elimination. These signaling pathways might be relevant to the pathogenesis of Alzheimer's disease.

Key words: postsynaptic density, glutamate receptor, dendritic spine, PSD-95, mass spectrometry, long term depression, synapse elimination

Introduction

Neurons in the mammalian central nervous system (CNS) communicate with each other via synapses. The vast majority of excitatory synapses occur at contacts between presynaptic axons and postsynaptic dendrites (often on specialized protrusions called dendritic spines), and they use glutamate as the neurotransmitter. Glutamate activates postsynaptic glutamate receptor-channels [primarily AMPA receptors (AMPARs) and NMDA receptors (NMDARs)], which open to depolarize the postsynaptic cell. NMDARs also allow influx of Ca^{2+} ions, which act as a second messenger to elicit biochemical changes in the postsynaptic neuron.

Synaptic plasticity is an umbrella term that describes multiple processes by which the strength of synaptic transmission can be modified by experience, specifically by previous synaptic activity (Malenka and Bear 2004). Synaptic plasticity allows synapses and circuits to store information, thereby contributing to mechanisms of learning and memory that can be very long-lasting. In addition to adjusting the strength of synaptic

[1] The Picower Institute for Learning and Memory, Howard Hughes Medical Institute, Depts of Brain and Cognitive Sciences, and Biology, Massachusetts Institute of Technology, Cambridge, MA, USA
Correspondence to: Morgan Sheng, The Picower Institute for Learning and Memory, Massachusetts Institute of Technology 77, Massachusetts Avenue (46-4303), Cambridge, MA 02139, USA, 617.452.3716 (tel), 617.452.3692 (fax), msheng@mit.edu

Selkoe et al.
Synaptic Plasticity and the Mechanism of Alzheimer's Disease
© Springer-Verlag Berlin Heidelberg 2008

transmission, neurons can change the morphology of synaptic connections (Yuste and Bonhoeffer 2001). Long-term information can be stored in the brain not only by modulation of synaptic strength but also by the formation of new synapses or the elimination of existing synapses (Chklovskii et al. 2004; Yuste and Bonhoeffer 2001). The elimination of many synapses occurs during normal development of the brain, particularly in postnatal stages of maturation. Developmental elimination of synapses is believed to be largely activity-dependent and accompanied by pruning of axons and dendrites to eliminate inappropriate connections (Cowan et al. 1984; Goda and Davis 2003; Lichtman and Colman 2000; O'Leary and Koester 1993; Weimann et al. 1999). The rates of formation and elimination of synapses are much less pronounced in adult brain. The key signaling events that regulate this physiologic retraction of dendrites/axons and elimination of synapses remain poorly understood.

Synaptic transmission and plasticity are crucial for all aspects of nervous system function and are critical for proper development of the CNS. To understand synaptic function and plasticity at the mechanistic molecular level, it is important to know the protein composition of synapses, how synaptic proteins are organized together as signaling complexes, the specific biochemical and cell biological functions of synaptic proteins, and how the abundance and activity of synaptic proteins are modified by activity. A bottoms-up approach has the potential to reveal unexpected mechanisms of synaptic function and plasticity because it is not predicated on physiological concepts.

The interest in the molecular mechanisms of synaptic function and plasticity has risen dramatically with the realization that major brain disorders, from neurodevelopmental illnesses to neurodegenerative diseases, may arise from dysfunction of synapses. In the case of Alzheimer's disease (AD), there is a growing belief that synapse dysfunction and synapse loss represent early steps in the pathology of the dementia (Walsh and Selkoe 2004). With regard to synaptic plasticity, recent evidence points to A-beta peptide having an inhibitory effect on synaptic transmission (Kamenetz et al. 2003; Walsh et al. 2002), with mechanisms overlapping with those of long-term depression (LTD; Hsieh et al. 2006).

I review here the molecular and cellular architecture of the postsynaptic specialization of glutamatergic (glutamate-releasing) synapses, which make up the vast majority of central synapses in mammalian brain. I focus on the postsynaptic density (PSD), where glutamate receptors are concentrated and where many of the signaling events that underlie synaptic plasticity occur. Some mechanisms that contribute to synapse loss and synaptic depression are particularly emphasized because of their relevance to AD.

Dendritic Spines

Excitatory synapses are usually situated at the tip of dendritic spines (typically 0.5–2 μm in length) that protrude from dendrite shafts. Spines occur at a density of up to 10 spines per μm of dendrite length on principal neurons (Sorra and Harris 2000) and they receive most of the excitatory synapses in the mature mammalian brain. Dendritic spines are highly heterogeneous structures that show dynamic motility, especially during development (Ethell and Pasquale 2005; Tada and Sheng 2006). Their number, size and shape undergo plastic changes correlated with long-term modifications of synaptic strength and inter-neuronal connectivity (Hayashi and Majewska

2005; Yuste and Bonhoeffer 2001). Many neurological and psychiatric diseases, including AD, are associated with altered morphology or loss of spines (Fiala et al. 2002).

The size of spine heads seems to increase with stimuli that strengthen synapses (e.g., long-term potentiation [LTP]) and decreases with stimuli that induce LTD (Hayashi and Majewska 2005; Kasai et al. 2003). The molecular mechanisms that coordinate synaptic strength with spine morphogenesis is a subject of current interest (Tada and Sheng 2006). Spines with large heads are generally stable, express large numbers of AMPARs, and contribute to strong synaptic connections. By contrast, spines with small heads are more motile, less stable, and contribute to weak synaptic connections (Holtmaat et al. 2006; Matsuzaki et al. 2004). Therefore, it is likely that synaptic depression is linked to morphological shrinkage and physical loss of synapses.

The Postsynaptic Density

The PSD is an electron-dense thickening of the postsynaptic membrane that lies directly apposed to the active zone. Usually located on the dilated tip (or head) of the spine, the PSD can be considered a huge membrane-associated protein complex specialized for postsynaptic signaling and plasticity (Kasai et al. 2003; Kennedy 2000; Sheng and Kim 2002; Siekevitz 1985). In addition to ionotropic glutamate receptors, the PSD contains a multitude of receptor tyrosine kinases, G-protein coupled receptors, ion channels and cell adhesion molecules, which together mediate physical linkage and/or functional communication with the presynaptic specialization, as well as postsynaptic signaling. These membrane receptors are linked to cytoplasmic scaffold proteins, signaling enzymes and cytoskeletal elements. Together the protein network is visible by electron microscopy as an irregular structure, often disk-like, of ~200–800 nm (mean 300–400 nm) width and ~30–50 nm thickness (Carlin et al. 1980).

Biochemical Composition and Signaling Pathways of the PSD

The composition of the PSD has been comprehensively investigated by biochemical purification followed by mass spectrometry, revealing hundreds of different proteins (Apperson et al. 1996; Cheng et al. 2006; Cho et al. 1992; Collins et al. 2005 2006; Dosemeci et al. 2006; Husi and Grant 2001; Husi et al. 2000; Jordan et al. 2004; Peng et al. 2004; Satoh et al. 2002; Walikonis et al. 2000; Walsh and Kuruc 1992; Yoshimura et al. 2004). Many PSD proteins were also indirectly identified by the yeast two-hybrid system as binding partners of known postsynaptic proteins, such as the identification of PSD-95 via its interaction with NR2 subunits of NMDARs (Funke et al. 2005; Kim and Sheng 2004; see Fig. 1). Identified proteins in the PSD include cell surface receptors, cytoplasmic signaling enzymes, and cytoskeletal and scaffold proteins (Fig. 1). Many of these proteins belong to small GTPase signaling pathways – especially Ras, Rho, Rac, Rap and Arf GTPses and their guanine nucleotide exchange factors (GEFs) and GTPase activating proteins (GAPs; Jordan et al. 2004; Peng et al. 2004; Fig. 1).

Rho family GTPases (especially RhoA, Rac1, Cdc42) are widely involved in control of the actin cytoskeleton. Actin is the main cytoskeleton of dendritic spines, and accordingly, the Rho family of GTPases is deeply implicated in regulation of dendritic spine morphogenesis (Govek et al. 2005; Tada and Sheng 2006). In particular, activation of RhoA, one of the best known Rho subfamily GTPases, has a strongly inhibitory effect on spine morphogenesis and causes profound loss of spines (Govek et al. 2005). However, little is known about the function of Rho during synaptic plasticity. Ras has long been believed to play a positive role in NMDAR signaling and in LTP and spine growth (Kennedy et al. 2005). One of the most abundant proteins in the PSD is SynGAP (see below), a negative regulator of Ras. It is possible that SynGAP plays an important inhibitory role in synapse development or maintenance.

Here we focus on Rap because it is particularly implicated in synaptic depression and loss of synapses. The small GTPase Rap1 and Rap2 are closely related members of the Ras GTPase family (~60% identity). Like Ras, Rap1 and Rap2 likely function upstream of MAP kinase cascades as well as other signaling pathways. For instance, Rap1 can stimulate Raf-ERK signaling (Stork 2003), and Rap2 can stimulate c-Jun NH2-terminal kinase (JNK) activity in non-neuronal cells (Machida et al. 2004).

Rap1 has been shown to be critical for LTD of synaptic transmission in hippocampal synapses (Huang et al. 2004; Morozov et al. 2003; Zhu et al. 2002) and causes inhibition of glutamatergic synaptic transmission (Imamura et al. 2003). Rap2, on the other hand, is required for synaptic depotentiation rather than LTD (Zhu et al. 2005). Consistent with a role in synaptic depression, activated Rap1 and Rap2 both inhibit surface expression of AMPARs in cultured hippocampal neurons (Fu et al. 2007). In addition, Rap2 induces retraction of axons and dendrites as well as the loss of spines and excitatory synapses. Interestingly, this effect of active Rap2 is seen in excitatory neurons but not in inhibitory interneurons (Fu et al. 2007). NMDA receptor stimulation in cultured cortical neurons induces activation of the small GTPase Rap1 (Xie et al. 2005). Activated Rap1 recruits the PDZ domain-containing protein AF-6 to the plasma membrane and induces spine neck elongation and smaller heads with reduced AMPAR content. In contrast, inactive Rap1 dissociates AF-6 from the membrane and induces growth of spine heads, associated with more AMPARs (Xie et al. 2005). In general, activated Rap promotes the weakening and shrinkage and possibly the loss of spines and synapses. In this sense, Rap seems to act antagonistically to Ras at excitatory synapses.

Proteomic screens and other experimental approaches have identified several GEFs and GAPs for Rap in the PSD (Apperson et al. 1996; Cheng et al. 2006; Cho et al. 1992;

Fig. 1. Major proteins and protein–protein interactions in the PSD, highlighting regulators of Ras and Rap. Schematic diagram of the network of proteins in the PSD. Major PSD proteins are shown in approximate stoichiometric ratio and scaled to molecular size, if known. Domain structure is shown only for PSD-95 (*red oval*, PDZ domain; *green*, SH3 domain; *blue*, guanylate kinase domain). CaMKII is depicted as dodecamer (*orange*). Un-named proteins in *gray* signify the other PSD proteins that are not illustrated in this diagram. Abbreviations: CAM, cell adhesion molecule; H, Homer; IRSp53, insulin receptor substrate 53; KCh, K^+ channel; mGluR, metabotropic glutamate receptor; RTK, receptor tyrosine kinase (e.g., ErbB4, TrkB); SPAR, spine associated RapGAP; SynGAP (synaptic Ras GAP). Ras promotes the growth and potentiation of excitatory synapses, whereas Rap promotes the weakening and shrinkage of synapses. SPAR inhibits the activity of Rap, and SynGAP inhibits the activity of Ras. The protein kinase Plk2 phosphorylates SPAR and induces its degradation via the ubiquitin-proteasome pathway

Collins et al. 2005a, b; Dosemeci et al. 2006; Husi and Grant 2001; Husi et al. 2000; Jordan et al. 2004; Peng et al. 2004; Satoh et al. 2002; Walikonis et al. 2000; Walsh and Kuruc 1992; Yoshimura et al. 2004). Consistent with a role for Rap in weakening and

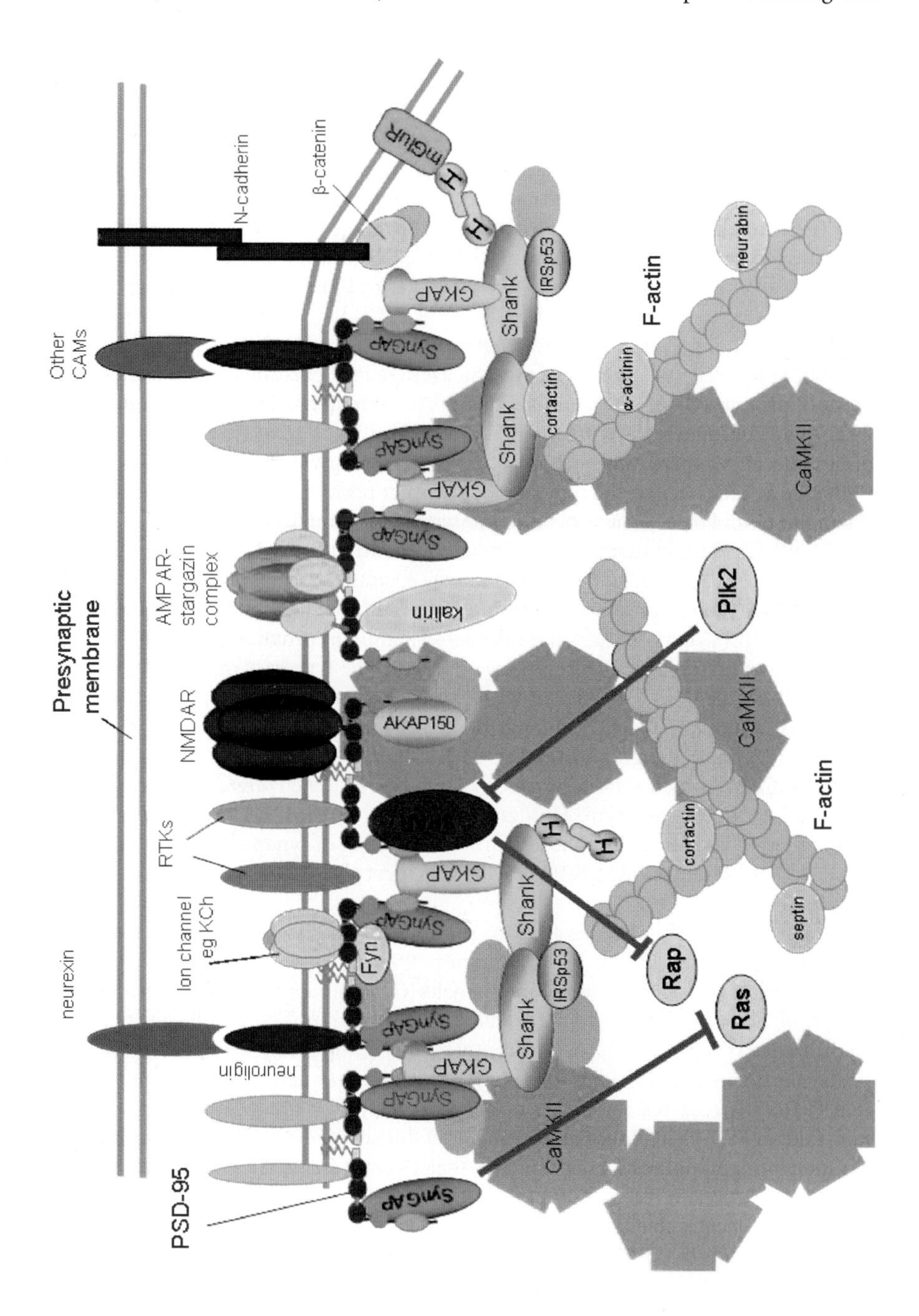

destabilization of synapses, growth of spines is enhanced by SPAR or Rap1GAP, two postsynaptic GAPs that inhibit Rap activity (Pak and Sheng 2003; Pak et al. 2001; Xie et al. 2005). The function of PSD RapGEFs, some of which are interestingly regulated by cAMP, is poorly understood and deserves greater attention (Bos 2006).

Quantitative View of PSD Protein Composition

Recently, a combination of quantitative electron microscopy and quantitative MS were used to determine the size, molecular weight and protein stoichiometry of an average PSD (Chen et al. 2005; Petersen et al. 2003). An "average PSD" of 360 nm diameter was estimated to have a total molecular mass of ~1.1 GDa, which would be equivalent to ~10 000 copies of a 100 kDa protein, or 100 copies of 100 different proteins of 100 kDa molecular weight, in an average PSD.

An important calibration of stoichiometry is provided by PSD-95, the best known scaffold of the PSD. We calculated that the average PSD contained ~300 copies of the scaffold protein PSD-95 (Chen et al. 2005). This number is in close agreement with Okabe and colleagues, who used a GFP-based optical method to estimate copy number of proteins at synapses (Sugiyama et al. 2005). Knowing the absolute stoichiometry of PSD-95, one can calculate the copy number of other proteins in the PSD by determining the molar abundance of the protein of interest relative to PSD-95.

Peng and colleagues used AQUA ("absolute quantitation" based on known amounts of added internal standards) MS to measure the absolute molar abundance of numerous PSD proteins (Cheng et al. 2006; Peng et al. 2004). CaMKIIα (28 pmol/20 μg PSD protein) and CaMKIIβ (4.7 pmol/20 μg) were the most abundant proteins in the PSD fraction purified from rat forebrain, representing ~7.4% and 1.3% mass of the PSD, respectively (Cheng et al. 2006), and consistent with earlier estimates (Kennedy et al. 1983). The next most abundant proteins among those measured were SynGAP, a postsynaptic RasGAP (2.1 pmol/20 μg), and the the PSD-95 family of scaffold proteins (to which SynGAP binds), which total ~2.3 pmol/20 μg (Fig. 1). It remains a mystery why two enzymes – CaMKII and SynGAP – are so plentiful in the PSD. It is interesting that one of these (CaMKIIα) is implicated in strengthening of synapses, whereas the other (SynGAP, which inhibits Ras-ERK activation; Kim et al. 2005) is involved in weakening of synapses.

PSD-95 itself, which binds to NR2 subunits of NMDA receptors, was found to be highly abundant in the adult forebrain PSD (1.7 pmol/20 μg of PSD protein; ~1% by mass) and more abundant than its closest relatives PSD-93/chapsyn-110 (0.3 pmol/20 μg) and SAP102 (0.2 pmol/20 μg). A wealth of data indicates that PSD-95 promotes synaptic strength, apparently by recruiting more AMPA receptors to the postsynaptic membrane (Elias et al. 2006). On the other hand, there is also evidence that PSD-95 is required for LTD, since overexpression of PSD-95 enhances LTD and knockout of PSD-95 in mouse promotes LTP at the expense of LTD (Migaud et al. 1998). An interesting hypothesis to consider is that PSD-95 enhances synaptic strength by recruiting glutamate receptors, but it promotes LTD by acting as a scaffold for those signaling molecules that are involved in LTD.

PSD-95 binds to GKAP/SAPAP, which interacts with Shank, which in turn binds to Homer (Kim and Sheng 2004; Fig. 1). These families of scaffold/adaptor proteins are abundant in the PSD. GKAP/SAPAP family proteins are about 30–40% as abundant as PSD-95 family proteins and approximately equimolar with Shank family proteins, and twice as abundant as Homer family proteins (Cheng et al. 2006). Assuming 300 copies of PSD-95 in an average PSD (Chen et al. 2005), an axis of scaffold proteins runs through the PSD with a stoichiometry of ~400 PSD-95 family: ~150 GKAP/SAPAP family: ~150 Shank family; ~60 Homer family (Cheng et al. 2006). Sugiyama et al. (2005) obtained higher values for Shank family proteins (~310 copies) and Homer proteins (~340) per average synapse, based on the GFP calibration approach in cultured neurons. The differences could be explained by differences in methodology and between brain tissue and neuronal culture. Despite these quantitative discrepancies, the PSD-95/GKAP/Shank/Homer scaffold proteins represent a substantial proportion of total protein mass of the PSD (Sugiyama et al. 2005). Here it should also be stressed that PSDs vary greatly in size in vivo, correlating with synaptic strength and abundance of glutamate receptors. Presumably the numbers of the above scaffolds are higher in large PSDs and lower in small PSDs.

NMDARs are the calcium-permeable glutamate receptors of the PSD and they play a central role in synaptic plasticity, being required for both LTP and LTD (Malenka and Bear 2004; Malinow and Malenka 2002). NMDARs are tightly associated with signaling and scaffold proteins and can be considered the functional core of the PSD. Based on the AQUA-MS method, we calculate that the number of NMDARs per average PSD is only ~20 (Cheng et al. 2006), roughly consistent with estimates based on electrophysiological approaches (Nimchinsky et al. 2004). NR2A and NR2B are both present in the NMDARs of the PSD. The number of NMDARs is far outnumbered (~20 fold) by the number of PSD-95 scaffolds to which NR2 subunits bind.

Quantitative MS counted ~60 copies of AMPAR subunits (GluR1, GluR2, GluR3) in the average PSD (Cheng et al. 2006), which corresponds to ~15 tetrameric AMPA receptor-channels. The majority of these (~80%) appears to be GluR1/GluR2 heteromers. The amount of AMPARs is likely to be an underestimate because approximately half of AMPARs is extracted by Triton X-100 during purification of PSDs.

Of course, biochemical approaches reveal only the *average* protein content of synapses. For instance, two PSDs containing 10 AMPA receptors each would give the same average as one containing 20 and another zero. It is known that the abundance of AMPAR is highly heterogeneous between individual synapses. It should also be emphasized that PSD composition varies greatly between different brain regions (e.g. forebrain versus cerebellum; Cheng et al. 2006). Protein composition also differs according to cell type within the same brain region, such that the PSD of hippocampal excitatory (glutamate-releasing) neurons have a distinctive protein make-up compared with PSDs of hippocampal inhibitory (GABA-releasing) neurons (Zhang et al. 1999). It will be important to investigate the structural and functional differences in excitatory postsynaptic signaling between glutamate- and GABA-releasing neurons.

Turnover and degradation of PSD proteins

Beyond their molecular complexity, one of the surprising discoveries about the PSD is the unexpectedly dynamic turnover of its constituent proteins. The composition of the PSD is modified by synaptic activity over a time scale of a few minutes to many hours, particularly by protein degradation via the ubiquitin-proteasome system (UPS; Ehlers 2003; Pak and Sheng 2003) and translocation of specific proteins (e.g., CaMKIIα and AMPARs) to and away from the PSD (Dosemeci et al. 2001; Hu et al, 1998; Inoue and Okabe 2003; Malinow and Malenka 2002). As measured by biochemical or imaging approaches, growing numbers of PSD proteins show a surprisingly high rate of turnover or mobility, including even "structural" proteins like actin and the PSD-95 family of scaffolds (Ehlers 2003; Inoue and Okabe 2003; Nakagawa et al. 2004; Star et al. 2002). Not only AMPARs but even PSD-95, which are frequently regarded as a stable scaffold of the PSD, show redistribution to neighboring synapses in the steady-state with a relatively fast time course (Borgdorff and Choquet 2002; Triller and Choquet 2005; Tsuriel et al. 2006). It will be interesting to investigate whether the stability of scaffold proteins such as PSD-95 within the PSD is regulated by activity through phophorylation and ubiqutination. Clearly the loss of synapses, as occurs in AD, might be presaged by increased turnover and accelerated degradation of PSD proteins.

Ubiquitination and activity-dependent degradation are reported for several PSD constituents, including the scaffolds PSD-95, GKAP and Shank (Colledge et al. 2003; Ehlers 2003). The proteasome appears to translocate towards postsynaptic sites in response to synaptic stimulation (Bingol and Schuman 2006), and proteasome inhibitors can block LTP and LTD (Colledge et al. 2003; Karpova et al. 2006). Thus degradation of PSD proteins by the UPS sculpts PSD structure and plays a major role in synaptic plasticity and presumably synapse turnover. It will be important to determine the nature of the specific E3 ubiquitin ligases that regulate the turnover of PSD proteins by the UPS, because they may be suitable targets for treatment in the synapse loss of neurodegeneration.

A striking example of regulated UPS-dependent degradation is that of SPAR (Spine Associated RapGAP), a GTPase activating protein for Rap that is enriched in the PSD and that interacts with PSD-95 (Pak and Sheng 2003; Pak et al. 2001). SPAR promotes the growth of dendritic spines, the postsynaptic compartment of excitatory synapses, at least in part by inhibiting postsynaptic Rap signaling (Pak et al. 2001). Bear in mind that Rap signaling seems to counteract Ras and lead to diminution of dendritic spines and weakening of synapses (Fu et al. 2007; Xie et al. 2005; Zhu et al. 2005); thus SPAR can be considered a promoter of synapse growth (Pak et al. 2001). Degradation of SPAR is induced by the protein kinase Polo-like kinase 2 [Plk2; also known as serum-inducible kinase (SNK), a member of the polo family of serine/threonine protein kinases. Plk2 levels are induced at the mRNA and protein levels in post-mitotic neurons by synaptic activity on the timescale of hours (Kauselmann et al. 1999; Pak and Sheng 2003). Plk3 (sometimes called FNK) is also expressed in postmitotic neurons and is inducible by activity (Kauselmann et al. 1999). Phosphorylation of SPAR by Plk2 leads to degradation of SPAR by the UPS, although the E3 ubiquitin ligase specifically required for this process remains to be identified. Plk2 itself downregulates synapses, presumably in part by inducing the loss of SPAR, as its overexpression causes depletion of mature mushroom spines and the overgrowth of thin, filopodia-like spines (Pak and Sheng 2003). The mechanism of action of Plk2 and its mode of specific recruitment to SPAR needs to be worked out. In any case, the degradation of SPAR by activity-inducible Plk2 provides a negative feedback mechanism for the weakening of synapses in neurons and could be involved in synaptic homeostasis.

Concluding Remarks

Recent years have seen explosive growth in our understanding of the structure and function of synapses, on both the pre- and postsynaptic sides. It seems likely that A-beta peptide exerts its negative influence, at least in part, on postsynaptic mechanisms of synaptic function and plasticity. Knowledge of the protein architecture of the PSD will facilitate identification of the target of A-beta peptide (the elusive "A-beta receptor"). Overall, it is important to remember that the PSD architecture is fluid, changeable, and predisposed for activity-regulated rearrangement; these dynamic properties are hidden by traditional biochemical or EM studies. The concept of the PSD as a dynamic structure in constant molecular flux raises fresh ways for thinking about the pathogenesis and treatment of chronic neurodegenerative diseases such as AD that take years to manifest pathology and symptoms.

Acknowledgements. M.S. is Investigator of the Howard Hughes Medical Institute. Work in M.S. lab is supported by NIH, and RIKEN-MIT Neuroscience Research Center.

References

Apperson ML, Moon IS, Kennedy MB (1996) Characterization of densin-180, a new brain-specific synaptic protein of the O-sialoglycoprotein family. J Neurosci 16:6839–6852.

Bingol B, Schuman EM (2006) Activity-dependent dynamics and sequestration of proteasomes in dendritic spines. Nature 441:1144–1148.

Borgdorff AJ, Choquet D (2002) Regulation of AMPA receptor lateral movements. Nature 417:649–653.

Bos JL (2006) Epac proteins: multi-purpose cAMP targets. Trends Biochem Sci 31:680–686.

Carlin RK, Grab DJ, Cohen RS, Siekevitz P (1980) Isolation and characterization of postsynaptic densities from various brain regions: enrichment of different types of postsynaptic densities. J Cell Biol 86:831–845.

Chen X, Vinade L, Leapman RD, Petersen JD, Nakagawa T, Phillips TM, Sheng M, Reese TS (2005) Mass of the postsynaptic density and enumeration of three key molecules. Proc Natl Acad Sci USA 102:11551–11556.

Cheng D, Hoogenraad CC, Rush J, Ramm E, Schlager MA, Duong DM, Xu P, Wijayawardana SR, Hanfelt J, Nakagawa T, Sheng M, Peng J (2006) Relative and absolute quantification of post-synaptic density proteome isolated from rat forebrain and cerebellum. Mol Cell Proteomics 5:1158–1170.

Chklovskii DB, Mel BW, Svoboda K (2004) Cortical rewiring and information storage. Nature 431:782–788.

Cho KO, Hunt CA, Kennedy MB (1992) The rat brain postsynaptic density fraction contains a homolog of the Drosophila discs-large tumor suppressor protein. Neuron 9:929–942.

Colledge M, Snyder EM, Crozier RA, Soderling JA, Jin Y, Langeberg LK, Lu H, Bear MF, Scott JD (2003) Ubiquitination regulates PSD-95 degradation and AMPA receptor surface expression. Neuron 40:595–607.

Collins MO, Yu L, Coba MP, Husi H, Campuzano I, Blackstock WP, Choudhary JS, Grant SG (2005) Proteomic analysis of in vivo phosphorylated synaptic proteins. J Biol Chem 280:5972–5982.

Collins MO, Husi H, Yu L, Brandon JM, Anderson CN, Blackstock WP, Choudhary JS, Grant SG (2006) Molecular characterization and comparison of the components and multiprotein complexes in the postsynaptic proteome. J Neurochem 97 Suppl 1:16–23

Cowan WM, Fawcett JW, O'Leary DD, Stanfield BB (1984) Regressive events in neurogenesis. Science 225:1258–1265.

Dosemeci A, Tao-Cheng JH, Vinade L, Winters CA, Pozzo-Miller L, Reese TS (2001) Glutamate-induced transient modification of the postsynaptic density. Proc Natl Acad Sci USA 98:10428–10432.

Dosemeci A, Tao-Cheng JH, Vinade L, Jaffe H (2006) Preparation of postsynaptic density fraction from hippocampal slices and proteomic analysis. Biochem Biophys Res Commun 339:687–694.

Ehlers MD (2003) Activity level controls postsynaptic composition and signaling via the ubiquitin-proteasome system. Nature Neurosci 6:231–242.

Elias GM, Funke L, Stein V, Grant SG, Bredt DS, Nicoll RA (2006) Synapse-specific and developmentally regulated targeting of AMPA receptors by a family of MAGUK scaffolding proteins. Neuron 52:307–320.

Ethell IM, Pasquale EB (2005) Molecular mechanisms of dendritic spine development and remodeling. Prog Neurobiol 75:161–205.

Fiala JC, Spacek J, Harris KM (2002) Dendritic spine pathology: cause or consequence of neurological disorders? Brain Res Brain Res Rev 39:29–54.

Fu Z, Lee SH, Simonetta A, Hansen J, Sheng M, Pak DT (2007) Differential roles of Rap1 and Rap2 small GTPases in neurite retraction and synapse elimination in hippocampal spiny neurons. J Neurochem 100:118–131.

Funke L, Dakoji S, Bredt DS (2005) Membrane-associated guanylate kinases regulate adhesion and plasticity at cell junctions. Annu Rev Biochem 74:219–245.

Goda Y, Davis GW (2003) Mechanisms of synapse assembly and disassembly. Neuron 40:243–264.

Govek EE, Newey SE, Van Aelst L (2005) The role of the Rho GTPases in neuronal development. Genes Dev 19:1–49.

Hayashi Y, Majewska AK (2005) Dendritic spine geometry: functional implication and regulation. Neuron 46:529–532.

Holtmaat A, Wilbrecht L, Knott GW, Welker E, Svoboda K (2006) Experience-dependent and cell-type-specific spine growth in the neocortex. Nature 441:979–983.

Hsieh H, Boehm J, Sato C, Iwatsubo T, Tomita T, Sisodia S, Malinow R (2006) AMPAR removal underlies Abeta-induced synaptic depression and dendritic spine loss. Neuron 52:831–843.

Hu BR, Park M, Martone ME, Fischer WH, Ellisman MH, Zivin JA (1998) Assembly of proteins to postsynaptic densities after transient cerebral ischemia. J Neurosci 18:625–633.

Huang CC, You JL, Wu MY, Hsu KS (2004) Rap1-induced p38 mitogen-activated protein kinase activation facilitates AMPA receptor trafficking via the GDI.Rab5 complex. Potential role in (S)-3,5-dihydroxyphenylglycene-induced long term depression. J Biol Chem 279:12286–12292.

Husi H, Grant SG (2001) Proteomics of the nervous system. Trends Neurosci 24:259–266.

Husi H,Ward MA, Choudhary JS, Blackstock WP, Grant SG (2000) Proteomic analysis of NMDA receptor-adhesion protein signaling complexes. Nature Neurosci 3:661–669.

Imamura Y, Matsumoto N, Kondo S, Kitayama H, Noda M (2003) Possible involvement of Rap1 and Ras in glutamatergic synaptic transmission. Neuroreport 14:1203–1207.

Inoue A, Okabe S (2003) The dynamic organization of postsynaptic proteins: translocating molecules regulate synaptic function. Curr Opin Neurobiol 13:332–340.

Jordan BA, Fernholz BD, Boussac M, Xu C, Grigorean G, Ziff EB, Neubert TA (2004) Identification and verification of novel rodent postsynaptic density proteins. Mol Cell Proteomics 3:857–871.

Kamenetz F, Tomita T, Hsieh H, Seabrook G, Borchelt D, Iwatsubo T, Sisodia S, Malinow R (2003) APP processing and synaptic function. Neuron 37:925–937.

Karpova A, Mikhaylova M, Thomas U, Knopfel T, Behnisch T (2006) Involvement of protein synthesis and degradation in long-term potentiation of Schaffer collateral CA1 synapses. J Neurosci 26:4949–4955.

Kasai H, Matsuzaki M, Noguchi J, Yasumatsu N, Nakahara H (2003) Structure-stability-function relationships of dendritic spines. Trends Neurosci 26:360–368.

Kauselmann G, Weiler M, Wulff P, Jessberger S, Konietzko U, Scafidi J, Staubli U, Bereiter-Hahn J, Strebhardt K, Kuhl D (1999) The polo-like protein kinases Fnk and Snk associate with a Ca(2+)- and integrin-binding protein and are regulated dynamically with synaptic plasticity. Embo J 18:5528–5539.

Kennedy MB (2000) Signal-processing machines at the postsynaptic density. Science 290:750–754.

Kennedy MB, Bennett MK, Erondu NE (1983) Biochemical and immunochemical evidence that the "major postsynaptic density protein" is a subunit of a calmodulin-dependent protein kinase. Proc Natl Acad Sci USA 80:7357–7361.

Kennedy MB, Beale HC, Carlisle HJ, Washburn LR (2005) Integration of biochemical signalling in spines. Nature Rev Neurosci 6:423–434.

Kim E, Sheng M (2004) PDZ domain proteins of synapses. Nature Rev Neurosci 5:771–781.

Kim MJ, Dunah AW, Wang YT, Sheng M (2005) Differential roles of NR2A- and NR2B-containing NMDA receptors in Ras-ERK signaling and AMPA receptor trafficking. Neuron 46:745–760.

Lichtman JW, Colman H (2000) Synapse elimination and indelible memory. Neuron 25:269–278.

Machida N, Umikawa M, Takei K, Sakima N, Myagmar BE, Taira K, Uezato H, Ogawa Y, Kariya KI (2004) Mitogen-activated protein kinase kinase kinase kinase 4 as a putative effector of Rap2 to activate the c-Jun N-terminal kinase. J Biol Chem 279:15711–15714

Malenka RC, Bear MF (2004) LTP and LTD: an embarrassment of riches. Neuron 44:5–21.

Malinow R, Malenka RC (2002) AMPA receptor trafficking and synaptic plasticity. Annu Rev Neurosci 25:103–126.

Matsuzaki M, Honkura N, Ellis-Davies GC, Kasai H (2004) Structural basis of long-term potentiation in single dendritic spines. Nature 429:761–766.

Migaud M, Charlesworth P, Dempster M, Webster LC, Watabe AM, Makhinson M, He Y, Ramsay MF, Morris RG, Morrison JH, O'Dell TJ, Grant SG (1998) Enhanced long-term potentiation and impaired learning in mice with mutant postsynaptic density-95 protein. Nature 396:433–439.

Morozov A, Muzzio IA, Bourtchouladze R, Van-Strien N, Lapidus K, Yin D, Winder DG, Adams JP, Sweatt JD, Kandel ER (2003) Rap1 couples cAMP signaling to a distinct pool of p42/44MAPK regulating excitability, synaptic plasticity, learning, and memory. Neuron 39:309–325.

Nakagawa T, Futai K, Lashuel HA, Lo I, Okamoto K, Walz T, Hayashi Y, Sheng M (2004) Quaternary structure, protein dynamics, and synaptic function of SAP97 controlled by L27 domain interactions. Neuron 44:453–467.

Nimchinsky EA, Yasuda R, Oertner TG, Svoboda K (2004) The number of glutamate receptors opened by synaptic stimulation in single hippocampal spines. J Neurosci 24:2054–2064.

O'Leary DD, Koester SE (1993) Development of projection neuron types, axon pathways, and patterned connections of the mammalian cortex. Neuron 10:991–1006.

Pak DT, Sheng M (2003) Targeted protein degradation and synapse remodeling by an inducible protein kinase. Science 302:1368–1373.

Pak DT, Yang S, Rudolph-Correia S, Kim E, Sheng M (2001) Regulation of dendritic spine morphology by SPAR, a PSD-95-associated RapGAP. Neuron 31:289–303.

Peng J, Kim MJ, Cheng D, Duong DM, Gygi SP, Sheng M (2004) Semiquantitative proteomic analysis of rat forebrain postsynaptic density fractions by mass spectrometry. J Biol Chem 279:21003–21011.

Petersen JD, Chen X, Vinade L, Dosemeci A, Lisman JE, Reese TS (2003) Distribution of postsynaptic density (PSD)-95 and Ca2+/calmodulin-dependent protein kinase II at the PSD. J Neurosci 23:11270–11278.

Satoh K, Takeuchi M, Oda Y, Deguchi-Tawarada M, Sakamoto Y, Matsubara K, Nagasu T, Takai Y (2002) Identification of activity-regulated proteins in the postsynaptic density fraction. Genes Cells 7:187–197.

Sheng M, Kim MJ (2002) Postsynaptic signaling and plasticity mechanisms. Science 298:776–780.

Siekevitz P (1985) The postsynaptic density: a possible role in long-lasting effects in the central nervous system. Proc Natl Acad Sci USA 82:3494–3498.

Sorra KE, Harris KM (2000) Overview on the structure, composition, function, development, and plasticity of hippocampal dendritic spines. Hippocampus 10:501–511.

Star EN, Kwiatkowski DJ, Murthy VN (2002) Rapid turnover of actin in dendritic spines and its regulation by activity. Nature Neurosci 5:239–246.

Stork PJ (2003) Does Rap1 deserve a bad Rap? Trends Biochem Sci 28:267–275.

Sugiyama Y, Kawabata I, Sobue K, Okabe S (2005) Determination of absolute protein numbers in single synapses by a GFP-based calibration technique. Nature Meth 2:677–684.

Tada T, Sheng M (2006) Molecular mechanisms of dendritic spine morphogenesis. Curr Opin Neurobiol 16:95–101.

Triller A, Choquet D (2005) Surface trafficking of receptors between synaptic and extrasynaptic membranes: and yet they do move! Trends Neurosci 28:133–139.

Tsuriel S, Geva R, Zamorano P, Dresbach T, Boeckers T, Gundelfinger ED, Garner CC, Ziv NE (2006) Local sharing as a predominant determinant of synaptic matrix molecular dynamics. PLoS Biol 4:e271.

Walikonis RS, Jensen ON, Mann M, Provance DW Jr., Mercer JA, Kennedy MB (2000) Identification of proteins in the postsynaptic density fraction by mass spectrometry. J Neurosci 20:4069–4080.

Walsh DM, Selkoe DJ (2004) Deciphering the molecular basis of memory failure in Alzheimer's disease. Neuron 44:181–193.

Walsh DM, Klyubin I, Fadeeva JV, Cullen WK, Anwyl R, Wolfe MS, Rowan MJ, Selkoe DJ (2002) Naturally secreted oligomers of amyloid beta protein potently inhibit hippocampal long-term potentiation in vivo. Nature 416:535–539.

Walsh MJ, Kuruc N (1992) The postsynaptic density: constituent and associated proteins characterized by electrophoresis, immunoblotting, and peptide sequencing. J Neurochem 59:667–678.

Weimann JM, Zhang YA, Levin ME, Devine WP, Brulet P, McConnell SK (1999) Cortical neurons require Otx1 for the refinement of exuberant axonal projections to subcortical targets. Neuron 24:819–831.

Xie Z, Huganir RL, Penzes P (2005) Activity-dependent dendritic spine structural plasticity is regulated by small GTPase Rap1 and its target AF-6. Neuron 48:605–618.

Yoshimura Y, Yamauchi Y, Shinkawa T, Taoka M, Donai H, Takahashi N, Isobe T, Yamauchi T (2004) Molecular constituents of the postsynaptic density fraction revealed by proteomic analysis using multidimensional liquid chromatography-tandem mass spectrometry. J Neurochem 88:759–768.

Yuste R, Bonhoeffer T. (2001) Morphological changes in dendritic spines associated with long-term synaptic plasticity. Annu Rev Neurosci 24:1071–1089.

Zhang W, Vazquez L, Apperson M, Kennedy MB, (1999) Citron binds to PSD-95 at glutamatergic synapses on inhibitory neurons in the hippocampus. J Neurosci 19:96–108.

Zhu JJ, Qin Y, Zhao M, Van Aelst L, Malinow R (2002) Ras and Rap control AMPA receptor trafficking during synaptic plasticity. Cell 110:443–455.

Zhu Y, Pak D, Qin Y, McCormack SG, Kim MJ, Baumgart JP, Velamoor V, Auberson YP, Osten P, van Aelst, L, Sheng M, Zhu JJ (2005). Rap2-JNK removes synaptic AMPA receptors during depotentiation. Neuron 46:905–916.

Impact of Beta Amyloid on Excitatory Synaptic Transmission and Plasticity

Roberto Malinow[1,2,3], *Helen Hsieh*[2], and *Wei Wei*[3]

Summary. There is a keen interest in identifying the effects of *Aβ* on synapses. Here we review some of the published work on its effects on excitatory synaptic transmission and plasticity. We also provide new information indicating that activity-induced release of *Aβ* from either pre- or postsynaptic neurons can block structural plasticity.

Introduction

A growing body of evidence implicates *Aβ* peptides and other derivatives of the amyloid precursor protein (APP) as being central to the pathogenesis of Alzheimer's disease (AD; Selkoe 2001; Hardy and Selkoe 2002). However, the functional relationship between APP and *Aβ* to neuronal electrophysiological function is poorly understood. There is also little information regarding the site, pre- or postsynaptic, from which *Aβ* originates in the generation of synaptic effects. Here we provide a selective review of studies that have addressed this issue and provide some preliminary data concerning the site from which *Aβ* originates.

In this short review we will focus on the effect of APP and *Aβ* on excitatory synaptic function, as there is considerable evidence that synapses are a key target site of AD (Terry et al. 1991). Two aspects of synaptic function affected by *Aβ* can be considered: "basal" synaptic transmission and synaptic plasticity. Of course, if synaptic plasticity is altered, and if one believes that the plasticity examined in experimental conditions occurs normally in the animal, one would expect "basal" transmission also to be altered. It is surprising how little this concept is considered in studies that examine the effect of any genetic perturbation on synaptic plasticity.

The effect of *Aβ* on synaptic transmission and plasticity can be examined using several experimental protocols, and it is not clear if results from different protocols should be directly compared. Furthermore, it must be kept in mind that ultimately one would like to make inferences with respect to the human condition, which obviously is a chronic condition that cannot be easily mimicked in animal experiments. Nevertheless, here we will consider results from such different protocols. Essentially there are three methods one can use to determine the effect of *Aβ* on synaptic function: 1) direct application of synthesized or purified *Aβ* on neural tissue; 2) examination of transgenic animals that over-express mutant forms of APP that generate increased amounts of *Aβ*;

[1] Cold Spring Harbor Laboratory, Cold Spring Harbor, New York 11724, USA, email: malinow@cshl.edu
[2] State University of New York and Stonybrook, Dept. Neurobiology
[3] Watson School of Biological Sciences, Cold Spring Harbor Laboratory

Selkoe et al.
Synaptic Plasticity and the Mechanism of Alzheimer's Disease
© Springer-Verlag Berlin Heidelberg 2008

or 3) transient over-expression of APP in neurons that drives increased amounts of *A*β. Each method has advantages and disadvantages. With the first method, the advantage is that the identity (namely *A*β) and its concentration are known. However, the oligomerizaton state of *A*β must be carefully controlled and it is not clear which oligomerized form is active or if there are some oligomerized forms that may be antagonistic to other active forms. Furthermore, with injection of *A*β into the brain, it is difficult to control the concentration of *A*β relative to the synapses being examined. With the second method, using transgenic mice, the concentrations of *A*β that are produced can be fairly well determined and, of course, the effects on behavior can be assessed. However, APP is chronically over-expressed and thus there may be compensatory mechanisms that obscure determining the direct effects of *A*β on synaptic function. Furthermore, different genetic strains may produce different amounts of *A*β, forming potentially different oligomeric forms and in different genetic backgrounds, which may complicate interpretations of results. With the third method, transient overexpression, the effects of various mutant forms of APP or *A*β can be compared, and the effects are relatively acute. However, levels of *A*β reaching synapses may not be easy to assess.

Effect of *A*β on Basal Excitatory Transmission

Early studies on the anatomy and immunohistochemistry of brains from AD patients indicated a dramatic loss of synapses and synaptic markers that correlated well with disease severity (Terry et al. 1991). Such markers were better correlated than brain plaque content to disease severity. The effect of *A*β on synaptic function was more carefully examined in a transgenic mouse model by Hsia et al. (1999). These animals, which express PDAPP(717V-F), showed considerable decrease in basal levels of excitatory synaptic function before the appearance of plaques, supporting the view that *A*β has effects on synaptic transmission independent of plaque formation. Similar findings using these mice were reported by Larson et al. (1999). However, a decrease in basal transmission has not been observed in all transgenic mouse strains. Studies employing mice derived from the Tg2576 APP(Sw) line have reported normal basal transmission in a slice preparation (Chapman et al. 1999) as well as in an in vivo preparation (Stern et al. 2004). It is not clear if the differences in these results are due to different levels or types of *A*β (e.g., 1-40 vs 1-42; different oligomeric forms) produced by these transgenic animals.

The effect on synapses following direct application of *A*β on neural tissue has been mixed. In some studies, enhancement of the NMDA component of transmission was found (Wu et al. 1995; Molnar et al. 2004); other studies showed no effect on basal transmission (Wang et al. 2004), whereas more recent results have measured a decrease in surface NMDA receptors (NMDA-Rs; Snyder et al. 2005). Part of this complexity may arise from the fact that increased activation of synaptic NMDA-Rs can lead to their removal (Morishita et al. 2005). It is not clear if application in slice preparations has the same effect as on dissociated cultured neurons. It is also not clear if the levels applied to tissue (even if they correspond to levels measured in extracellular space of AD patients) are those sensed by synapses during the disease process. It is possible that synaptic regions close to focal sources of *A*β will be exposed to much higher levels of *A*β than those levels measured in the extracellular space. Why might there be locally high sources of $A\beta$ during the disease process? Senile plaques, which contain large amounts of $A\beta s$ may release their contents to the nearby extracellular space. Thus, sites close to plaques may be exposed to high levels of $A\beta$.

An alternative means to determine the effect of APP and its proteolytic products on synapses is to transiently express APP in neurons by acute gene transfer, which can be achieved by viral or biolistic delivery techniques. One advantage of this technique is that different APP-related constructs can be delivered and compared. Furthermore, other constructs can be co-expressed along with APP, allowing one to dissect mechanisms by which APP and its proteolytic products act. We have used acute gene delivery techniques to express APP and related constructs in individual neurons of organotypic hippocampal slices (Kamenetz et al. 2003; Hsieh et al. 2006). Transfected neurons can be identified by the co-expressed GFP. Simultaneous whole-cell patch clamp recordings are obtained from transfected and nearby non-transfected neurons. By evoking synaptic transmission with a stimulus electrode, transmission onto these two neurons can be compared, and thus the effect of expression of a given construct on synaptic transmission can be determined.

We found that expression of APP for 1 day produced a depression of excitatory synaptic transmission with no effect on inhibitory transmission, neuronal input resistance or resting potential. These findings indicate that the effect on excitatory transmission was not due to a general toxic effect on the neuron. The effect could be narrowed down to formation of $A\beta$ since 1) expression of APP with a point mutation that prevents cleavage by beta secretase (APP_{MV}) does not produce synaptic depression, 2) expression of the beta secretase product of APP (C99) is sufficient to produce synaptic depression and 3) expression of C99 in the presence of gamma secretase inhibitors produces no synaptic depression. More recent results indicate that the synaptic depression produced by $A\beta$ is due to synaptic removal of AMPA-Rs (Hsieh et al. 2006). For instance, co-expression of an AMPA-R subunit with a mutation that reduces its endocytosis blocks the synaptic depressing effect of APP. Interestingly, expression of this AMPA-R subunit also prevents the depression of NMDA-R-mediated transmission (Hsieh et al. 2006), suggesting that $A\beta$-driven endocytosis of NMDA-R (Snyder et al. 2005) requires endocytosis of AMPA-Rs.

The synaptic depressing effects of $A\beta$ are interesting when coupled with findings indicating that increased neural activity can drive the processing of APP to $A\beta$ (Kamenetz et al. 2003). These two findings suggest a negative feedback system: high levels of neural activity drive formation of $A\beta$; formation of $A\beta$ depresses synaptic transmission and thereby reduces high levels of neural activity. We have hypothesized that the $A\beta$-mediated negative feedback system is a normal process that may become dysregulated (e.g., by producing unregulated amounts of $A\beta$) and thereby contribute to AD.

Effect of $A\beta$ on Long-Term Potentiation

Long-term potentiation (LTP) is a persistent enhancement in synaptic transmission that follows a brief period of repeated pre- and postsynaptic activity (Bliss and Collingridge 1993). LTP is observed in many CNS synapses, is long-lasting, demonstrates associative properties and thus has served as a powerful model of synaptic plasticity that may underlie associative forms of learning and memory. It is therefore clear why many groups have tested the effects of AD-related molecules and model systems on LTP.

Most studies testing the effects of *A*β on LTP have found that this form of plasticity is reduced (Cullen et al. 1997; Lambert et al. 1998; Chapman et al. 1999; Moechars et al. 1999; Chen et al. 2002; Walsh et al. 2002; Kamenetz et al. 2003). It is notable, however, that a group with much expertise in LTP studies failed to see a deficit of LTP in a transgenic mouse overexpressing APP, despite clear reduction in synaptic transmission (Hsia et al. 1999).

It cannot be over-emphasized that testing the effects of *A*β on LTP is not as trivial as it may seem. When synthetic *A*β is put in solution it can aggregate, which will generate different products with potentially different and variable effects. As a small hydrophobic peptide, it can adhere to perfusion tubing, which will affect the concentration in solution. When *A*β is injected in vivo, it is very difficult to control its concentration, as sites close to the injection will have a high concentration whereas more distant sites will have a lower concentration; the volume over which the injected solution spreads is virtually impossible to measure. In one remarkable study (Walsh et al. 2002), injection of a few picograms of *A*β oligomers into brain ventricles was sufficient to block LTP. Rough calculations suggest that a few thousand molecules per neuron may be sufficient to block LTP.

Transgenic models have provided insight into the effects of *A*β on plasticity but these studies also have limitations. The levels of expression of APP and proteolytic products differ in different strains. Furthermore, background genetic make-up may modify the effects of *A*β. It is also important to note that effects of APP or proteolytic products can affect transmission, manifesting themselves as effects on LTP. For instance, if increased expression of APP produces a decrease in excitatory transmission, then one may see decreased LTP merely because there is not sufficient depolarization during a tetanic stimulus. In such a case, the effect is not directly on the biochemical pathways underlying LTP.

It is important to emphasize that physiology experiments should be done, as much as possible, in a blind manner, that is, the experimenter should be blind as to whether a test or control compound (genotype) is being tested, because there are many hidden biases that can creep into physiology experiments. For instance, it is not uncommon to remove from analysis a particular experiment because the recording was not stable. The "stability of a recording," however, is a somewhat subjective judgment. Thus, if one is blind during the experiment and during the analysis, these sorts of hidden biases can be avoided.

Is Pre- or Postsynaptic *A*β Responsible for Synaptic Effects?

Previous studies have suggested that APP is transported and released at presynaptic sites (Buxbaum et al. 1998; Lazarov et al. 2002). To address where the *A*β that has effects on synapses is released, we have recently examined the effects of *A*β on spine structural changes during LTP. We can induce spine structural changes by brief bath application of a solution that drives neurons in slices to fire in a burst pattern. This leads to LTP (termed chemical LTP, or cLTP; Kopec et al. 2006). During cLTP there is an increase in electrophysiological responses at synapses adn an increase in AMPA-Rs at synapses, and the potentiation is blocked by blockade of protein kinases (PKC, CaMKII) required for standard LTP (Kopec et al. 2006). There is also a robust increase in spine size (Kopec et al. 2006). In this new study, we find that expression of APP in postsynaptic neurons largely blocks the increase in spine size. We believe the blocking of plasticity is due to production of *A*β because the spine potentiation is rescued if a gamma secretase inhibitor is included in the bath and also the blocking of plasticity is mimicked by bath application of *A*β.

To test if *A*β secretion from one postsynaptic pyramidal neuron could affect another pyramidal neuron, we expressed GFP in ~50% on neurons with one virus, and in fewer neurons (~5%) we expressed dsRed along with APP. Thus red neurons were making APP (and *A*β) whereas green neurons were not making *A*β. We monitored spines on green neurons close to (< 3μm) or far from (> 10μm) a red dendrite before and after cLTP. We found that spines close to an APP-expressing cell had significantly attenuated structural changes, whereas spines far from such cells showed normal structural changes. These results indicate that spines close to cells overproducing Aβ had their structural plasticity blocked.

To test if presynaptic Aβ can affect postsynaptic structural plasticity, we infected presynaptic CA3 cells with a virus expressing APP and dsRed and postsynaptic cells with a virus expressing GFP. Again we monitored postsynaptic spines close to (<3 μm) or far from (>10 μm) a red presynaptic axon. We found that spines on cells not overexpressing APP, but close to presynaptic axons that do overexpress APP, had their structural plasticity reduced during cLTP. Spines that were far from APP-expressing axons, but in the same slice, showed normal structural plasticity. We conclude that Aβ generated from pre- or postsynaptic sites can block activity-induced structural plasticity.

References

Bliss TV, Collingridge GL (1993) A synaptic model of memory: long-term potentiation in the hippocampus. Nature 361:31–39.

Buxbaum JD, Thinakaran G, Koliatsos V, O'Callahan J, Slunt HH, Price DL, Sisodia SS (1998) Alzheimer amyloid protein precursor in the rat hippocampus: transport and processing through the perforant path. J Neurosci 18:9629–9637.

Chapman PF, White GL, Jones MW, Cooper-Blacketer D, Marshall VJ, Irizarry M, Younkin L, Good MA, Bliss TV, Hyman BT, Younkin SG, Hsiao KK (1999) Impaired synaptic plasticity and learning in aged amyloid precursor protein transgenic mice. Nature Neurosci 2:271–276.

Chen QS, Wei WZ, Shimahara T, Xie CW (2002) Alzheimer amyloid beta-peptide inhibits the late phase of long-term potentiation through calcineurin-dependent mechanisms in the hippocampal dentate gyrus. Neurobiol Learn Mem 77:354–371.

Cullen WK, Suh YH, Anwyl R, Rowan MJ (1997) Block of LTP in rat hippocampus in vivo by beta-amyloid precursor protein fragments. Neuroreport 8:3213–3217.

Hardy J, Selkoe DJ (2002) The amyloid hypothesis of Alzheimer's disease: progress and problems on the road to therapeutics. Science 297:353–356.

Hsia AY, Masliah E, McConlogue L, Yu GQ, Tatsuno G, Hu K, Kholodenko D, Malenka RC, Nicoll RA, Mucke L (1999) Plaque-independent disruption of neural circuits in Alzheimer's disease mouse models. Proc Natl Acad Sci USA 96:3228–3233.

Hsieh H, Boehm J, Sato C, Iwatsubo T, Tomita T, Sisodia S, Malinow R (2006) AMPAR removal underlies Abeta-induced synaptic depression and dendritic spine loss. Neuron 52:831–843.

Kamenetz F, Tomita T, Hsieh H, Seabrook G, Borchelt D, Iwatsubo T, Sisodia S, Malinow R (2003) APP processing and synaptic function. Neuron 37:925–937.

Kopec CD, Li B, Wei W, Boehm J, Malinow R (2006) Glutamate receptor exocytosis and spine enlargement during chemically induced long-term potentiation. J Neurosci 26:2000–2009.

Lambert MP, Barlow AK, Chromy BA, Edwards C, Freed R, Liosatos M, Morgan TE, Rozovsky I, Trommer B, Viola KL, Wals P, Zhang C, Finch CE, Krafft GA, Klein WL (1998) Diffusible, nonfibrillar ligands derived from Abeta1-42 are potent central nervous system neurotoxins. Proc Natl Acad Sci USA 95:6448–6453.

Larson J, Lynch G, Games D, Seubert P (1999) Alterations in synaptic transmission and long-term potentiation in hippocampal slices from young and aged PDAPP mice. Brain Res 840:23–35.

Lazarov O, Lee M, Peterson DA, Sisodia SS (2002) Evidence that synaptically released beta-amyloid accumulates as extracellular deposits in the hippocampus of transgenic mice. J Neurosci 22:9785–9793.

Moechars D, Dewachter I, Lorent K, Reverse D, Baekelandt V, Naidu A, Tesseur I, Spittaels K,Haute CV, Checler F, Godaux E, Cordell B, Van Leuven F (1999) Early phenotypic changes in transgenic mice that overexpress different mutants of amyloid precursor protein in brain. J Biol Chem 274:6483–6492.

Molnar Z, Soos K, Lengyel I, Penke B, Szegedi V, Budai D (2004) Enhancement of NMDA responses by beta-amyloid peptides in the hippocampus in vivo. Neuroreport 15:1649–1652.

Morishita W, Marie H, Malenka RC (2005) Distinct triggering and expression mechanisms underlie LTD of AMPA and NMDA synaptic responses. Nature Neurosci 8:1043–1050.

Selkoe DJ (2001) Alzheimer's disease: genes, proteins, and therapy. Physiol Rev 81(2):741–766.

Snyder EM, Nong Y, Almeida CG, Paul S, Moran T, Choi EY, Nairn AC, Salter MW, Lombroso PJ, Gouras GK, Greengard P (2005) Regulation of NMDA receptor trafficking by amyloid-beta. Nature Neurosci 8:1051–1058.

Stern EA, Bacskai BJ, Hickey GA, Attenello FJ, Lombardo JA, Hyman BT (2004) Cortical synaptic integration in vivo is disrupted by amyloid-beta plaques. J Neurosci 24:4535–4540.

Terry RD, Masliah E, Salmon DP, Butters N, DeTeresa R, Hill R, Hansen LA, Katzman R (1991) Physical basis of cognitive alterations in Alzheimer's disease: synapse loss is the major correlate of cognitive impairment. Ann Neurol 30:572–580.

Walsh DM, Klyubin I, Fadeeva JV, Cullen WK, Anwyl R, Wolfe MS, Rowan MJ, Selkoe DJ (2002) Naturally secreted oligomers of amyloid beta protein potently inhibit hippocampal long-term potentiation in vivo. Nature 416:535–539.

Wang Q, Rowan MJ, Anwyl R (2004) Beta-amyloid-mediated inhibition of NMDA receptor-dependent long-term potentiation induction involves activation of microglia and stimulation of inducible nitric oxide synthase and superoxide. J Neurosci 24:6049–6056.

Wu J, Anwyl R, Rowan MJ (1995) beta-Amyloid selectively augments NMDA receptor-mediated synaptic transmission in rat hippocampus. Neuroreport 6:2409–2413.

Quantitative Neuropathology in Alzheimer's Mouse Models

Floyd E. Bloom[1]

Summary. Transgenic (Tg) mouse models of Alzheimer's Disease (AD) provide insight into the earliest changes observable as potential strategic targets by which to mount therapeutic interventions. A comparison of the different mouse models to determine the earliest time at which diffuse or compact amyloid deposits are detected versus other structural pathology indicates that multiple factors combine to produce highly variable courses across Tg mouse models. Therefore, we sought to define the earliest time point at which alterations could be detected and whether such changes were progressive in two different mouse models, the Elan PDAPP mouse and the original Tg2576 mouse. In the Elan PDAPP mouse, Tg hippocampal volumes were statistically significantly smaller at 100 days but the volume differences did not progress. Furthermore, the volume loss was restricted to the dentate gyrus (DG), where the difference between Tg and wild-type (WT) brain was nearly 30% by 100 days. Amyloid deposition was not detected until six months of age. Subsequent studies confirmed that the outer molecular layer (OML) terminal projections of the lateral entorhinal cortex layer II neurons were the earliest and most consistent sites of synaptic dysfunction. With age, PDAPP WT mice eventually also showed loss of OML dendritic spines. In the Tg2576 mice, no hippocampal or cortical volumetric differences were detected, but OML dendritic spine analysis, electrophysiology and behavior were all affected (versus WT mice) by four months of age. Amyloid was not detectable until 15 months of age. Others have recently confirmed these observations on the dentate OML synapse loss in the older Tg2576 mouse. Thus, early neurotoxic fragments may make highly vulnerable synapses the earliest site pathology in these simulations of familial AD.

Introduction

Familial forms of Alzheimer's Disease (AD) offer a starting point to recapitulate the pathological processes in transgenic (Tg) mouse models of the disease. Because it is possible to define pathology early in the course of a Tg mouse's life history, before signs of functional deficit emerge, finding the earliest possible evidence of neuropathology can identify the most vulnerable neurons, suggest disease mechanisms, and provide a means to refine novel treatments based on the progression of biomarkers during a Tg mouse's life rather than on the end stage neuropathologies seen in human brains at post-mortem.

The classical method for defining cellular neuropathology has been microscopy. Over the past few years, high-throughput, rigorous and standardized methodologies have been developed for research on brain structures, from cells to macro-regions to whole brains, and have been complemented by protocols for analysis of gene expression patterns within the three-dimensional context of the brain's structure, circuits and cells

[1] Molecular and Integrative Neuroscience Department, The Scripps Research Institute, La Jolla CA 92037, USA, email: fbloom@scripps.edu

Selkoe et al.
Synaptic Plasticity and the Mechanism of Alzheimer's Disease
© Springer-Verlag Berlin Heidelberg 2008

(see Broide et al. 2004; Redwine et al. 2003; Reilly et al. 2003; Wu et al. 2004; Zapala et al. 2005).

AD has been found to have highly inheritable familial forms. Tg mouse models of the mutated forms of one likely AD candidate protein, the amyloid precursor protein (APP), have been achieved, alone or in combination with known mutations in other familial forms, including Pre-senilin and Tau (see Bloom et al. 2005 and references therein). Each of the mutated APPs showed amino acid substitutions around the proteolytic cleavage sites of the β or γ secretases. Therefore, the resulting abnormal fragments of APP, Aβ1-40 or 1-42, have been held to be the neurotoxic agents of the disease, and drugs developed to block these proteases or means to blunt their effects (e.g., absorption by antibodies) have been pursued as treatments (Schenk et al. 2001).

As one compares the different mouse models for the earliest time at which diffuse or compact amyloid deposits are detected, it is clear that the mutation, the transcriptional promoter used to drive its expression, the background strains in which the mutation is fostered, as well as several other incompletely known factors combine to produce highly variable courses across Tg mouse models. However, aside from the intracerebral accumulations of the diffuse or compact aggregates of Aβ, little other mouse neuropathology aside from loss of synaptic proteins has yet been described.

In addition, it is unclear when Tg mice begin to show the pathological effects of the transgene and when the optimal times of intervention might be, or whether the structural alterations precede or follow the behavioral dysfunctions. Therefore, we sought to define the earliest time point at which alterations could be detected, and whether such changes were progressive in two Tg mouse models of AD, the Elan PDAPP mouse (see Games et al. 2006 for review) and the commercially available Tg2576 mouse (Hsiao et al. 1996). We employed quantitative magnetic resonance microscopy (MRM) and stereology to compare the volumes of the hippocampus in the PDAPP Tg mice at different ages and in wild-type (WT) mice of the same strain (see Redwine et al. 2003). Although Tg and WT mice were statistically identical at 40 days, hippocampal volume was statistically significantly smaller by 12.3% in Tg vs. WT mice at 100 days and did not progress. Furthermore, the volume loss was restricted to the dentate gyrus (DG) where the difference between the Tg mouse brain and the WT brain was nearly 30% at the 100-day point.

We next applied our technologies to define quantitatively the spatial and temporal progression of the age-dependent accumulation of Aβ in the most vulnerable regions (see Reilly et al. 2003). Minimal deposition of Aβ was first detected at six months of age in PDAPP mice, primarily as thioflavin-positive, compact amyloid, widely dispersed in hippocampus and cortex, that increased only moderately with age. Diffuse amyloid immunostaining increased dramatically between 12 and 15 months in all subfields, particularly the DG. Amyloid load in the DG was significantly above all other hippocampal subregions at 15, 18, and 22 months. The Aβ appeared to be distributed in a lamina-specific pattern within the DG, highest in the inner and outer molecular layers (IML and OML, respectively). Inputs to the OML and MML derive from the lateral and medial entorhinal cortices, respectively, and input to the IML derives from the dentate hilus. Although total Aβ loads in the entorhinal cortex prior to 15 months were negligible, between 15 and 22 months, there was a progressive increase in diffuse Aβ deposition selectively in the lateral entorhinal cortex (reaching 16.4% of the volume), whereas loads in the medial entorhinal cortex remained below 2.3%, demonstrating a circuit-specific accumulation of Aβ (Reilly et al. 2003).

Subsequently, we pursued the cellular explanation for the volume reduction in the 90-day Tg mouse DG by reconstructing dentate granule cell (GC) dendritic complexity (Wu et al. 2004) with high-throughput diolistic cell loading and 3D neuronal reconstruction. In 90-day PDAPP mice, analysis of all sampled GC types revealed a 12% reduction of total dendritic length in PDAPP mice compared with WT littermate controls. Further analysis, performed with refined subgroups, found that superficially located GCs in the dorsal blade were most profoundly altered, exhibiting a 23% loss in total dendritic length, whereas neurons in the ventral blade were unaffected. Superficial GCs of the dorsal blade were particularly vulnerable (a 32% reduction) in the posterior region of the DG. Thus, substantial dendritic pathology is evident in 90-day PDAPP mice for a spatially defined subset of GCs well before amyloid accumulation occurs. Extending these studies to older PDAPP mice revealed that ventral blade GCs do later lose dendritic complexity as well by 15 months of age. While there is progressive loss of dendritic complexity in the PDAPP mice, by 15 months of age it only matches that seen in equal-age WT mice (Reilly, Wu and Bloom, unpublished observations).

In contrast, in the Tg2576 mouse model (originally termed APP_{695}SWE), amyloid plaques do not develop until 18 months (Hsiao et al. 1996), although both long-term potentiation (LTP) deficits in hippocampal CA1 and DG and spatial memory deficits in a modified water maze were detected at six months. These findings suggest impaired synaptic plasticity because there was no observable loss of presynaptic or postsynaptic structural elements or neurons (Chapman et al. 1999). The spatial memory deficits observed at six months were correlated with an elevation of detergent-insoluble Aß aggregates (Westerman et al. 2002). Although pre-plaque reduction in synaptophysin immunodetection has been reported in PDAPP mice at two to three months of age (Hsia et al. 1999), Tg2576 mice had not been studied in this regard.

We therefore performed a systematic evaluation, from behavior to synaptic connections, of the temporal progression of neuronal dysfunction in the Tg2576 mouse model of AD. We observed that significant deficits in synaptic connectivity, as revealed by loss of dendritic spines in the OML in Golgi-impregnated dentate granule cells and diminished contextual fear conditioned behavior occurred at four months, just prior to perforant path to dentate LTP deficits were observed at five months and before the measurable rise of insoluble Aß42 levels at six months (Jacobsen et al. 2006). In contrast, as previously noted, amyloid plaque deposition in these animals was not detectable until 12–18 months of age. Measuring spine densities in the molecular layer of the DG using a modified stereologic approach to count spines in Golgi-impregnated neurons, we demonstrated a significant reduction of spine densities at four months of age that coincides with the first evidence of impaired hippocampal learning in contextual fear conditioning (CFC). Tg animals examined at younger ages did not display synaptic deficits and were not impaired in the learning task. We concluded that a reduction in spine density correlating with CFC impairment is the earliest evidence of neuronal dysfunction observed in this mouse model of AD

The reduction in spine density seen in the OML of the DG of heterozygous Tg2576 mice is significant for two reasons. First, such rigorously quantified pathology at this age in this Tg model had not previously been reported. In principle, this pathologic marker could be exploited as a potential indicator of therapeutic interventions. Second, the reduced spine density is a sensitive neuroanatomical substrate for the impairments in memory and LTP seen in this amyloid-based mouse model of AD, effectively modeling the early memory defects seen in humans with AD.

The decrease in spine density observed in the OML of the young Tg2576 animals, and the changes in synaptic physiology within the same region, are consistent with the alterations in dendritic branching observed in the same brain region of the young PDAPP mice (Wu et al. 2004). These data suggest that dysfunction of the circuitry of the OML of the DG, including the afferent projects from the lateral entorhinal cortex, is one of the earliest neuroanatomical effects of APP overexpression. The loss of dendritic spines seen at four months of age shows progression when re-evaluated at 12 and 15 months of age, although at the latter time point, WT mice also begin to show OML spine loss. Recently, using ultrastructural analysis of synaptic bouton density in the OML of 6- to 15-month-old Tg2576 mice, the spine loss seen by Golgi stain analysis was confirmed (Dong et al. 2005). In addition to such murine data, studies of aging in rats and non-human primates suggest that this region is among the most sensitive to age-associated disruptions (Small et al. 2004).

Deficits that occur after the early-onset deficits discussed above include an increase in total amyloid load and in the number of reactive astrocytes and microglia. There is accumulating evidence that molecular, morphological, functional, and behavioral deficits can be measured for extended periods preceding the first evidence of plaque formation. In Tg2576 animals, these changes occur at approximately four to five months of age and include a significant increase in the fraction of Aß42 to Aß40. The long time interval between synaptic, electrophysiological, and behavioral deficits and plaque accumulation and gliosis was not expected and is important for understanding the course of disease and the development of therapeutic strategies beneficial to early treatment. Hence, the presence of one or more species of soluble Aß, containing an elevated proportion of Aß42, may trigger some of the early-onset morphological and functional synaptic deficits that lead to memory dysfunction in this model.

These data highlight the importance that quantitative neuropathological examination may have in the treatment of AD.

Acknowledgements. The author acknowledges the diligent contributions of Drs. John Morrison, Jeff Redwine, John Reilly, Chi-Cheng Wu, and Warren Young to the data and protocols developed at Neurome Inc, 2000–2006 described in this report.

References

Bloom FE, Reilly JF, Redwine JM, Wu CC, Young WG, Morrison JH (2005) Mouse models of human neurodegenerative disorders: requirements for medication development. Arch Neurol 62:185–187.

Broide RS, Trembleau A, Ellison JA, Cooper J, Lo D, Young WG, Morrison JH, Bloom FE (2004) Standardized quantitative in situ hybridization using radioactive oligonucleotide probes for detecting relative levels of mRNA transcripts verified by real-time PCR. Brain Res 1000:211–222.

Chapman PF, White GL, Jones MW, Cooper-Blacketer D, Marshall VJ, Irizarry M, Younkin L, Good MA, Bliss TV, Hyman BT, Younkin SG, Hsiao KK (1999) Impaired synaptic plasticity and learning in aged amyloid precursor protein transgenic mice. Nature Neurosci 2:271–276.

Dong H, Csernansky CA, Martin MV, Bertchume A, Vallera D, Csernansky JG (2005) Acetylcholinesterase inhibitors ameliorate behavioral deficits in the Tg2576 mouse model of Alzheimer's disease. Psychopharmacology (Berl) 181:145–152.

Games D, Buttini M, Kobayashi D, Schenk D, Seubert P (2006) Mice as models: transgenic approaches and Alzheimer's disease. J Alzheimers Dis 9:133–149.

Hsia AY, Masliah E, McConlogue L, Yu GQ, Tatsuno G, Hu K, Kholodenko D, Malenka RC, Nicoll RA, Mucke L (1999) Plaque-independent disruption of neural circuits in Alzheimer's disease mouse models. Proc Natl Acad Sci USA 96:3228–3233.

Hsiao K, Chapman P, Nilsen S, Eckman C, Harigaya Y, Younkin S, Yang F, Cole G (1996) Correlative memory deficits, Abeta elevation, and amyloid plaques in transgenic mice [see comment]. Science 274:99–102.

Jacobsen JS, Wu CC, Redwine JM, Comery TA, Arias R, Bowlby M, Martone R, Morrison JH, Pangalos MN, Reinhart PH, Bloom FE (2006) Early-onset behavioral and synaptic deficits in a mouse model of Alzheimer's disease. Proc Natl Acad Sci USA 103:5161–5166.

Redwine JM, Kosofsky B, Jacobs RE, Games D, Reilly JF, Morrison JH, Young WG, Bloom FE (2003) Dentate gyrus volume is reduced before onset of plaque formation in PDAPP mice: a magnetic resonance microscopy and stereologic analysis. Proc Natl Acad Sci USA 100:1381–1386.

Reilly JF, Games D, Rydel RE, Freedman S, Schenk D, Young WG, Morrison JH, Bloom FE (2003) Amyloid deposition in the hippocampus and entorhinal cortex: quantitative analysis of a transgenic mouse model. Proc Natl Acad Sci USA 100:4837–4842.

Schenk D, Games D, Seubert P (2001) Potential treatment opportunities for Alzheimer's disease through inhibition of secretases and Abeta immunization. J Mol Neurosci 17:259–267.

Small SA, Chawla MK, Buonocore M, Rapp PR, Barnes CA (2004) Imaging correlates of brain function in monkeys and rats isolates a hippocampal subregion differentially vulnerable to aging. Proc Natl Acad Sci USA 101:7181–7186.

Westerman MA, Cooper-Blacketer D, Mariash A, Kotilinek L, Kawarabayashi T, Younkin LH, Carlson GA, Younkin SG, Ashe KH (2002) The relationship between Abeta and memory in the Tg2576 mouse model of Alzheimer's disease. J Neurosci 22:1858–1867.

Wu CC, Chawla F, Games D, Rydel RE, Freedman S, Schenk D, Young WG, Morrison JH, Bloom FE (2004) Selective vulnerability of dentate granule cells prior to amyloid deposition in PDAPP mice: digital morphometric analyses. Proc Natl Acad Sci USA 101:7141–7146.

Zapala MA, Hovatta I, Ellison JA, Wodicka L, Del Rio JA, Tennant R, Tynan W, Broide RS, Helton R, Stoveken BS, Winrow C, Lockhart DJ, Reilly JF, Young WG, Bloom FE, Lockhart DJ, Barlow C (2005) Adult mouse brain gene expression patterns bear an embryologic imprint. Proc Natl Acad Sci USA 102:10357–10362.

Multiple Levels of Synaptic Regulation by NMDA-type Glutamate Receptor in Normal and Disease States

Veronica A. Alvarez[1], *Ganesh M. Shankar*[1,2], *Brenda L. Bloodgood*[1], *Dennis J. Selkoe*[2], and *Bernardo L. Sabatini*[1]

Summary. The acquisition of new behaviors and the formation of memories occur through the creation and regulation of synaptic contacts within the brain. In mammals, most synapses form onto small, bulbous cellular compartments called dendritic spines (reviewed in Harris 1999). Spines are dynamic structures that appear rapidly following activity patterns that lead to memory formation, and these fast structural alterations are believed to contribute to the remarkable plasticity of the brain (Engert and Bonhoeffer 1999; Maletic-Savatic et al. 1999; Toni et al. 1999). Each spine is biochemically isolated (Sabatini et al. 2002) and contains components of many signaling pathways necessary for synaptic plasticity (Kornau et al. 1995). Here we describe recent work in our laboratory focusing on the role of NMDA-type glutamate receptors (NMDAR) in regulating the function and plasticity of dendritic spines and synapses in both normal and disease states (Alvarez et al.2007; Ngo-Anh et al. 2005; Bloodgood and Sabatini 2007a; Shankar et al. 2007).

Introduction

Neuronal activity regulates both synapse formation and elimination and hence influences the plasticity and the pruning of neuronal circuits (reviewed in Alvarez and Sabatini 2007). NMDA-type glutamate receptors (NMDARs) play a critical role in many forms of activity-dependent regulation of synapses demonstrated both in vitro and in vivo (Malenka and Nicoll 1993; Fox et al. 1996; Huang and Pallas 2001). For example, changes in sensory experience alter connectivity in sensory cortex by inducing synapse formation (Knott et al. 2002), accelerating the rate of spine elimination (Zuo et al. 2005), and inducing NMDAR-dependent plasticity (Sawtell et al. 2003). In vitro, patterned electrical stimulation can drive the NMDAR-dependent formation or elimination of dendritic protrusions in the hippocampus (Maletic-Savatic et al. 1999; Nagerl et al. 2004). In addition, NMDAR activity influences the efficacy of synapses by inducing the insertion or removal of AMPA-type glutamate receptors (AMPARs) from synapses during long-term potentiation (LTP) and depression (LTD), respectively. Thus, NMDARs play a central role in the rapid regulation of synaptic transmission but their contribution to the long-term stabilization of glutamatergic synapses is unknown.

NMDAR-mediated calcium (Ca) influx into dendritic spines is a key trigger of many forms of synaptic plasticity. Synaptically evoked calcium transients ($\Delta[\mathrm{Ca}]$) in spines are determined by the activation/inactivation kinetics and Ca permeability of

[1] Department of Neurobiology, Harvard Medical School, Boston MA 02115, email: bernardo_sabatini@hms.harvard.edu

[2] Center for Neurologic Diseases, Brigham and Women's Hospital and Harvard Medical School, Boston, MA 02115

Selkoe et al.
Synaptic Plasticity and the Mechanism of Alzheimer's Disease
© Springer-Verlag Berlin Heidelberg 2008

Ca sources, the concentration and affinity of endogenous Ca-binding proteins, the efficiency of Ca extrusion, and the morphology of the spine (reviewed in Bloodgood and Sabatini, 2007b). These factors, by governing the induction of long-term plasticity and the activation of Ca-dependent ion channels, determine the amplitude of synaptic potentials. However, determining the role of each of these factors in setting synaptically evoked Ca transients and potentials has been difficult due to the small size of the dendritic spine.

Here we describe recent work from our laboratory using optical and electrophysiological methods to examine the relationship between the NMDARs and synaptic plasticity. We find that NMDAR-mediated Ca transients within active spines are regulated by multiple ion channels, including SK-type Ca-activated K-channels (SK) and $Ca_V2.3$ voltage-sensitive Ca channels (VSCCs; Ngo-Anh et al. 2005; Bloodgood and Sabatini 2007a). Furthermore, we find that oligomers of the amyloid-β (Aβ) protein, a peptide that is central in the pathogenesis of Alzheimer's disease (AD), reduces NMDAR-dependent Ca influx in spines and triggers the loss of synapses in an NMDAR-dependent manner (Shankar et al. 2007). Lastly, we show that NMDARs also control the number of synapses and the density of dendritic spines through a pathway that is independent of receptor opening but requires protein-protein interactions mediated by the C-tail of the NR1 subunit of the receptor (Alvarez et al. 2007).

NMDAR-dependent Regulation of Synaptic Potentials and Calcium Transients

We have used 2-photon laser-scanning microscopy (2PLSM) combined with 2-photon laser uncaging (2PLU) of caged glutamate (Fig. 1) to directly stimulate glutamate receptors at the post-synaptic terminal on a visualized dendritic spine while monitoring intracellular Ca levels. This approach allows us to bypass the presynaptic terminal and record evoked signals in pharmacological conditions that prevent normal release of neurotransmitter from the presynaptic terminal (Carter and Sabatini 2004; Bloodgood and Sabatini 2007a). For example, the involvement of voltage-gated Ca and Na channels in shaping synaptic signals has been difficult to determine because activation of these channels is necessary for action potential (AP)-evoked presynaptic release of neurotransmitter. Thus combining conventional electrophysiological and pharmacological techniques to probe the involvement of these channels in postsynaptic signaling is not possible. Furthermore, some voltage-gated Ca channels are found within dendritic spines and their roles in locally shaping electrical and biochemical signals evoked by synaptic stimulation may not be detectable in somatic electrophysiological recordings.

We have found that $Ca_V2.3$ VSCCs are present in dendritic spines of mouse CA1 pyramidal neurons but absent from the adjacent dendrite (Bloodgood and Sabatini 2007a). Therefore, we hypothesized that this channel may play a role in regulating synaptic signaling and may couple selectively to SK-type Ca-activated K-channels known to be present in the spine head (Ngo-Anh et al. 2005). To test this idea, the EPSP evoked by 2PLU (uEPSP) and the Ca transients at the stimulated spines ($\Delta[\mathrm{Ca}]_{uEPSP}$) were measured in a variety of pharmacological conditions. In control conditions and at a recording temperature of 33 °C, a 500 μs laser pulse at a power level sufficient to

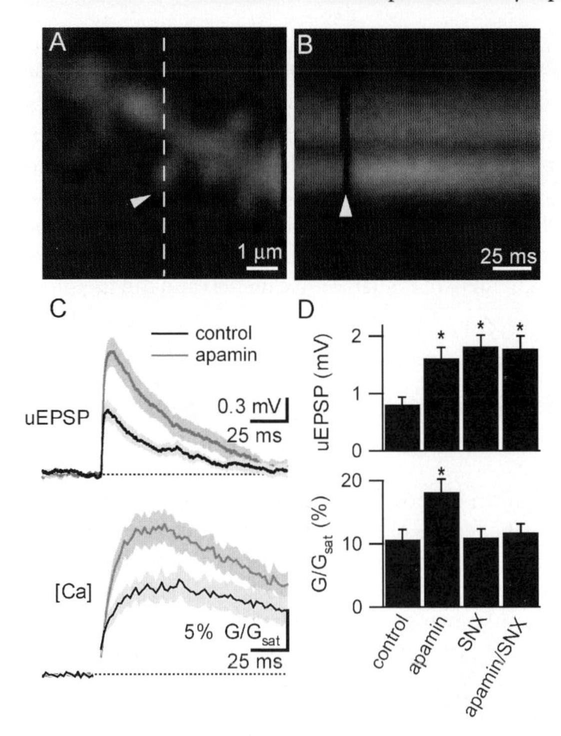

Fig. 1. Control of synaptic potentials and spine head calcium transients SK and $Ca_V2.3$ channels. **A.** Image of a spiny apical dendrite of a mouse CA1 pyramidal neuron filled through a whole-cell recordings electrode with Alexa 594 (*red fluorescence*) and the Ca indicator Fluo-5F (*green fluorescence*). MNI-glutamate, a caged version of glutamate, is included in the bath. **B.** Fluorescence collected in a line that intersects the spine head and the neighboring dendrite (indicated by the dashed line in panel A). The *arrowheads* in A and B indicate the location and timing of a 500µs pulse of 720 nm light that photoreleases the glutamate. **C.** Uncaging-evoked EPSPs (uEPSP, top) and spine head fluorescence transients ([Ca], *bottom*) recorded in control conditions (*black*) or in the presence of the SK channel antagonist apamin (*red*). The *lines* and *shaded areas* indicate the means and standard errors about the mean, respectively. **D.** Amplitude of the uEPSPs (*top*) and spine head fluorescence transients ($\Delta G/Gsat$, *bottom*) in control conditions, or in the presence of apamin, SNX-482, or apamin and SNX-482. $*p < 0.05$ compared with control

photobleach ~40% of the red fluorescence resulted in uEPSP = 0.81 ± 0.13 mV and $\Delta[\mathrm{Ca}]_{uEPSP} = 10.6 \pm 1.6\%\Delta G/G_{sat}$ (n = 12/6 spines/cells; Fig. 1A–D). In the presence of apamin, a peptide antagonist of SK channels, the uEPSP and $\Delta G/G_{sat}$ increased to 1.62 ± 0.19 mV and 18.1 ± 2.0%, respectively ($p < 0.05$; n = 12/4 spines/cells; Fig. 1C, D). Application of SNX-482, a selective blocker of $Ca_V2.3$ VSCCs, increased the uEPSP (1.81 ± 0.20 mV, $p < 0.5$) without affecting $\Delta[\mathrm{Ca}]_{uEPSP}$ ($\Delta G/G_{sat} = 11.0 \pm 1.3$; n = 14/9 spines/cells; Fig. 1D). Coapplication of SNX and apamin resulted in synaptic signals that resembled those seen in the presence of SNX alone, (uEPSP = 1.77 ± 0.23, $\Delta G/G_{sat} = 11.8 \pm 1.3$; n = 14/4 spines/cells; Fig. 1D), confirming that activation of $Ca_V2.3$ VSCCs is necessary for the effects of apamin on synaptic signals at near-physiological temperatures.

From these studies and a larger pharmacological analysis, we conclude that synaptic potentials recorded at the soma do not simply reflect the level of AMPA receptor activation at the synapse but are also influenced by the activity of voltage-sensitive Ca channels and Ca-activated K channels (Ngo-Anh et al. 2005; Bloodgood and Sabatini 2007a). Similarly, the amplitudes of synaptically evoked Ca transients in the spine head do not simply reflect the number of active NMDA receptors but are also dependent on the activity of glutamate-independent ion channels, such as $Ca_V2.3$, located in the spine head. Furthermore, Ca transients within the spine head and synaptic potentials are differentially dependent on the complement of available ion channels and can therefore be regulated independently.

NMDAR-dependent Loss of Spines and Excitatory Synapses Triggered by Aβ Oligomers

Although the causes of AD are unknown in most patients, amyloid-β (Aβ) protein is known to play a major role in the pathogenesis of the disease. This is clear since many familial forms of AD are caused by mutations in the β-amyloid precursor protein (APP) or in the presenilins, proteases that mediate the production of Aβ from APP (reviewed in Selkoe, 2001). Nevertheless, it is unknown which of the many forms of Aβ found in the brains of AD patients induce the neurological symptoms that characterize the disease. Furthermore, although rises in soluble Aβ in the cortex correlate with disease progression in AD patients and in animal models of the disease (Kuo et al. 1996; McLean et al. 1999; Mucke et al. 2000; Naslund et al. 2000; Moolman et al. 2004; Spires et al. 2004), no specific soluble form of Aβ that induces AD-like neuropathological changes has been identified. To determine which specific forms of soluble Aβ trigger synapse loss, we examined the effects of Aβ monomers and oligomers on excitatory synapses and dendritic spines in rat hippocampus.

Previous reports had suggested that synthetic forms of Aβ at micromolar concentrations resulted in a loss of surface NMDARs (Snyder et al. 2005). Because micromolar concentrations of synthetic Aβ allow the peptide to form multiple soluble and insoluble aggregates, we relied instead on naturally secreted Aβ that can be biochemically separated into monomer and low-n oligomer fractions (Podlisny et al. 1995; Walsh et al. 2000). These fractions are isolated by size-exclusion chromatography of media collected from Chinese hamster ovary cells that stably express human APP_{751} with the Val717Phe mutation (7PA2 cells; Podlisny et al., 1995). We found that acute application of fractions containing Aβ dimers and trimers to rat hippocampal brain slices resulted in a rapid reduction of uncaging-evoked, NMDAR-dependent Ca transients within spines (Fig. 2A, B). No effect was seen with Aβ monomers and neither fraction altered uEPSPs recorded at the soma.

To examine the long-term consequences of exposure to Aβ, we applied Aβ oligomers or monomers to organotypic slices of rat hippocampus and monitored dendritic spine density (Fig. 3). Slices were prepared from postnatal day 5–7 (P5–7) rats and biolistically transfected with green-fluorescence protein (GPF) at two days in vitro (DIV). At 5 DIV, fractions containing either Aβ monomers or Aβ oligomers were added to the slice culture medium (SCM) and replenished every two to three days. Aβ oligomers and monomers were stable in SCM and no loss or interconversion of Aβ isoforms was

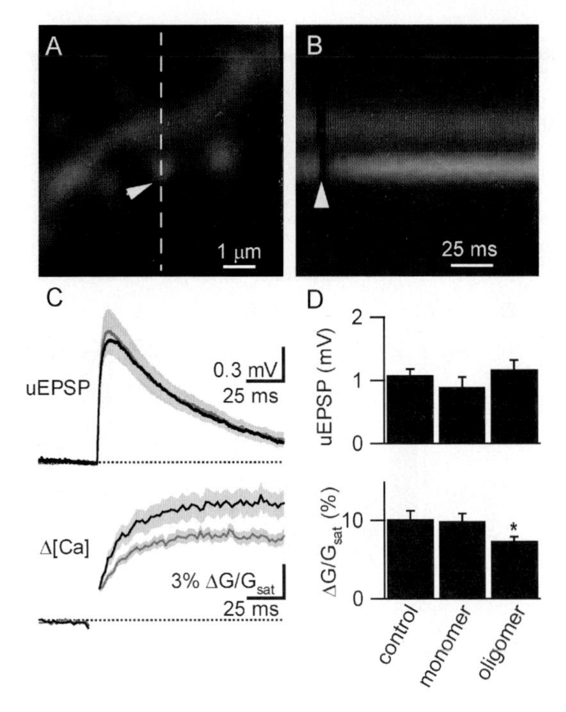

Fig. 2. Aβ oligomers decrease NMDAR-mediated Ca transients in spine heads stimulated with glutamate uncaging. **A.** Image of an apical dendrite of a rat CA1 pyramidal neuron filled with Alexa 594 (*red fluorescence*) and the Ca indicator Fluo-5F (*green fluorescence*). **B.** Fluorescence collected in a line that intersects the spine head and the neighboring dendrite (indicated by the *dashed line* in panel A). The *arrowheads* in A and B indicate the location and timing of the laser pulse used to uncage glutamate. **C.** uEPSPs (*top*) and spine head fluorescence transients (Δ[Ca], *bottom*) recorded in control conditions (*red*) or in the presence of oligomers of Aβ (*black*). The lines and shaded areas indicate the means and standard errors about the mean, respectively. **D.** Average amplitude of uEPSPs (*top*) and spine head fluorescence transients (ΔG/Gsat, *bottom*) in control conditions, or following an acute exposure to Aβ monomers or oligomers. $*p < 0.05$ compared with control

seen (Fig. 3A). After 15 days of treatment, neurons exposed to Aβ oligomers had severely decreased spine density compared to those exposed to monomers or left in control SCM (Fig. 3B). The effect on spine density was progressive and loss of ~50% of spines was visible after five days of exposure (Fig. 3C). Spine loss reflected a true loss of excitatory synapses, as Aβ oligomer exposure also triggered a decrease in the frequency of spontaneous miniature excitatory postsynaptic currents (mEPSCs) detected by whole-cell voltage-clamp recordings (data not shown; Shankar et al. 2007).

Spine loss was specifically caused by the Aβ in the fractions as it was prevented by coapplication of antibody 6E10, which was raised against human Aβ (Kim et al. 1988), but not by boiled, denatured 6E10 (Fig. 3D). Similarly, the active form of *scyllo*-inositol (AZD-103) (McLaurin et al. 2000), a small molecule that binds to Aβ and reduces its aggregation, prevented the effects of Aβ oligomers whereas the inactive enantioner, *chiro*-inositol, did not (Fig.3D). Lastly, Aβ oligomer-induced spine loss required active

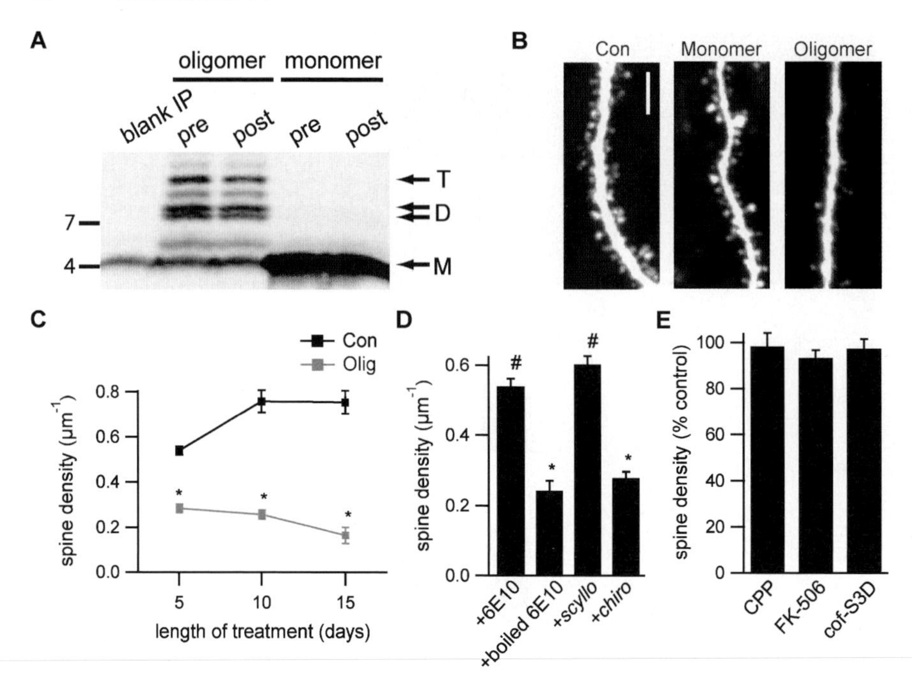

Fig. 3. Naturally secreted Aβ oligomers, but not monomers, decrease dendritic spine density in hippocampal pyramidal neurons through an NMDAR-dependent mechanism. **A.** Immunoprecipitation of Aβ with polyclonal antibody R1282 (Walsh et al. 2000) was performed on slice culture medium (SCM) containing reconstituted size exclusion chromatography (SEC) fractions of 7PA2 conditioned medium. Immunoprecipitation was done both before (pre) and after (post) a 2-day incubation with organotypic slices. The blank IP lane (far left) shows the results of a similar analysis of control SCM. **B.** Representative images of apical dendrites of pyramidal cells in organotypic hippocampal slice cultures treated with control SCM (Con), SCM with added Aβ monomers (Monomer), or SCM with added Aβ oligomers (Oligomer) 10 days. *Scale bar*: 5 µm. **C.** Summary of spine density (mean ±SEM) in apical dendrites of Con and Oligomer- (Olig) treated neurons as in (b) for 5, 10 or 15 days. $* = p < 0.05$ compared to Con. **D.** Slices were treated for five days with Aβ oligomers and the Aβ-specific antibody 6E10 (+6E10), Aβ oligomers and boiled 6E10 (+boiled 6E10), Aβ oligomers and *scyllo*-inositol (AZD-103) (+*scyllo*), or Aβ oligomers and the inactive steroisomer *chiro*-inositol (+*chiro*). Spine density is expressed as mean ± SEM. $*, \# = p < 0.05$ compared to Con or Olig, respectively, from C. **E.** Summary of spine density of slices treated with Aβ oligomers and 20 µM CPP for 10 days (CPP), Aβ oligomers and 1 µM FK-506 for 24 hours after a nine-day control treatment (FK-506), or Aβ oligomer treatment for 10 days of neurons expressing cofilin-S3D (cof-S3D). Spine density measured in the presence of each pharmacological agent and Aβ oligomers is shown relative to that measured in the presence of the pharmacological agent alone

NMDARs, calcineurin (protein phosphatase 2B), and the actin depolymerization factor cofilin, as application of Aβ oligomers had no effect on spine density of cells incubated in the NMDAR antagonist CPP, the calcineurin antagonist FK-506, or of neurons expressing a constitutively inactive mutant of cofilin (cof-S3D).

From these studies and a more extensive pharmacological analysis, we propose a model (Fig.4) in which Aβ oligomers partially inhibit NMDAR-mediated Ca sig-

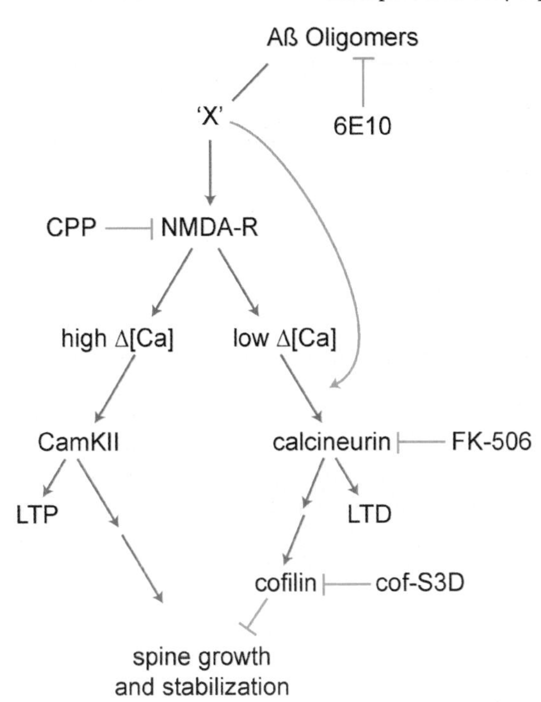

Fig. 4. Proposed pathways that regulate spine density and that are affected by Aβ oligomers. *Blue lines* indicate elements of the pathway tested in our studies and '*X*' indicates a postulated receptor that may bind Aβ and lead to inhibition of NMDARs-mediated synaptic Ca influx

naling and activate a pathway that favors the induction of LTD and spine loss. Ca influx through synaptic NMDARs can activate at least two pathways that regulate spine density. In the pathway on the left side of the figure, high levels of Ca induce LTP via a CAMKII-dependent pathway (reviewed in Cummings et al., 1996). LTP-inducing stimuli also trigger enlargement of dendritic spines and growth of new spines in a NMDAR- and CAMKII-dependent manner (Engert and Bonhoeffer 1999; Maletic-Savatic et al. 1999; Jourdain et al. 2003; Matsuzaki et al. 2004; Nagerl et al. 2004). In the pathway on the right, low levels of Ca, such as those reached during low-frequency subthreshold stimulation, induce LTD through a calcineurin dependent pathway (reviewed in Malenka and Bear 2004). LTD-inducing stimuli also lead to spine shrinkage via an NMDAR/calcineurin/cofilin-dependent pathway and spine retraction through an NMDAR-dependent pathway (Zhou et al. 2004; Nagerl et al. 2004). In this model, full blocking of NMDARs interrupts both pathways, leading to no net spine loss.

We found that Aβ oligomers decreased spine density in an NMDAR/calcineurin/cofilin-dependent manner, consistent with activation of the pathway shown on the right. Aβ oligomers may reduce NMDAR-dependent Ca transients such that stimuli normally activating the left hand pathway instead activate those on the right. This might occur through direct interaction of Aβ with NMDARs or by activating unknown factors ('X') that may lead to inhibition of NMDARs-mediated synaptic Ca influx. Our proposed model is largely consistent with a recent report that also concludes that Aβ decreases dendritic spine density and alters glutamatergic signalling in CA1 pyramidal neurons in rat hippocampal organotypic slices (Hsieh et al. 2006).

Ionotropic and Structural Roles of NMDA Receptors in Regulating Synapse Stability

To determine the ionotropic and structural roles of NMDARs in regulating synapse stability, we compared in rat organotypic hippocampal slices the effects of blocking NMDARs with pharmacological antagonists to those of physically eliminating the receptor from the synapse (Alvarez et al. 2007). To physically eliminate NMDARs from neurons, we used RNA-interference (RNAi) to target its NR1 subunit, which is necessary for assembly of functional receptors (McIlhinney et al. 2003; Schorge and Colquhoun 2003). RNAi was performed with vector-driven expression of small hairpin RNA that allows for long-term knock-down of the targeted protein. This vector, named shNR1, also expresses a fluorescent marker (GFP) that allows for the identification of the transfected neuron and analysis of its morphology (Tavazoie et al. 2005). Neurons expressing shNR1 exhibit reduced NR1 immunostaining in the soma compared to neurons expressing a control vector encoding only GFP after 10 days of transfection. In addition, recordings from control and shNR1 neurons in hippocampal organotypic cultures show the synaptic currents evoked by extracellular stimulation of Shaffer collaterals and show a reduced NMDAR-mediated synaptic component in shNR1 neurons (data not shown). Depolarization to +40 mV of control neurons relieves the Mg block of NMDARs and reveals a long-lived current that is sensitive to the NMDAR antagonist CPP (10 M). shNR1-transfected neurons display faster-decaying outward currents at +40 mV, and the ratio between the long-lasting current at −40 mV (NMDAR-mediated current) and the peak of the current at −60 mV (AMPAR-mediated current), $R_{+40/-60}$, is reduced in shNR1 neurons ($R_{+40/60} = 0.29 \pm 0.04$, 0.08 ± 0.01, and 0.04 ± 0.02 for control, shNR1, and CPP neurons, respectively; $n = 10, 12, 3$). Thus, at 10 days post transfection (DPT), shNR1-expressing neurons display a severe reduction of NMDAR-mediated synaptic currents but still receive functional synapses.

Pyramidal neurons in organotypic cultures undergo synaptic maturation that parallels many aspects of neuronal development in vivo and are, therefore, well-suited for studying the role of the NMDARs in the longer-term stabilization of spines and synapses (Fig. 5). During the first three weeks in vitro, the density of spines in control pyramidal neurons increased from $0.5 \pm 0.02\ \mu m^{-1}$ at 10 DPT to $0.8 \pm 0.04\ \mu m^{-1}$ and $0.93 \pm 0.03\ \mu m^{-1}$ at 15 and 20 DPT, respectively ($n = 12-25$ cells). Pyramidal neurons expressing shNR1 showed reduced spine density at 10 DPT ($0.35 \pm 0.02\ \mu m^{-1}$), which declined to reach $0.18 \pm 0.02\ \mu m^{-1}$ and $0.12 \pm 0.01\ \mu m^{-1}$ at 15 and 20 DPT, respectively (Fig. 5A).

The frequency of mEPSCs was significantly reduced in shNR1 neurons relative to controls (0.5 ± 0.1 Hz and 0.11 ± 0.03 Hz in control and shNR1 neurons, respectively; $n = 15-18$), without changes in the amplitude of the individual events (control = -21 ± 0.9 pA; shNR1 = -20 ± 1.3 pA; $n = 15-18$; Alvarez et al. 2007). Thus, postsynaptic expression of NMDARs is required for the stabilization of dendritic spines and the maintenance of functional excitatory synapses.

To determine if both the ionotropic and protein interaction components of NMDAR signaling are necessary to maintain spine density, the effects of the NMDAR antagonist D-2-Amino-5-phosphonovaleric acid (AP5, 50 μM) on spine density was examined (Fig. 5A). Neurons in organotypic slice cultures were transfected with GFP and chronically treated with AP5. Long-term blockade of NMDARs did not trigger spine elimi-

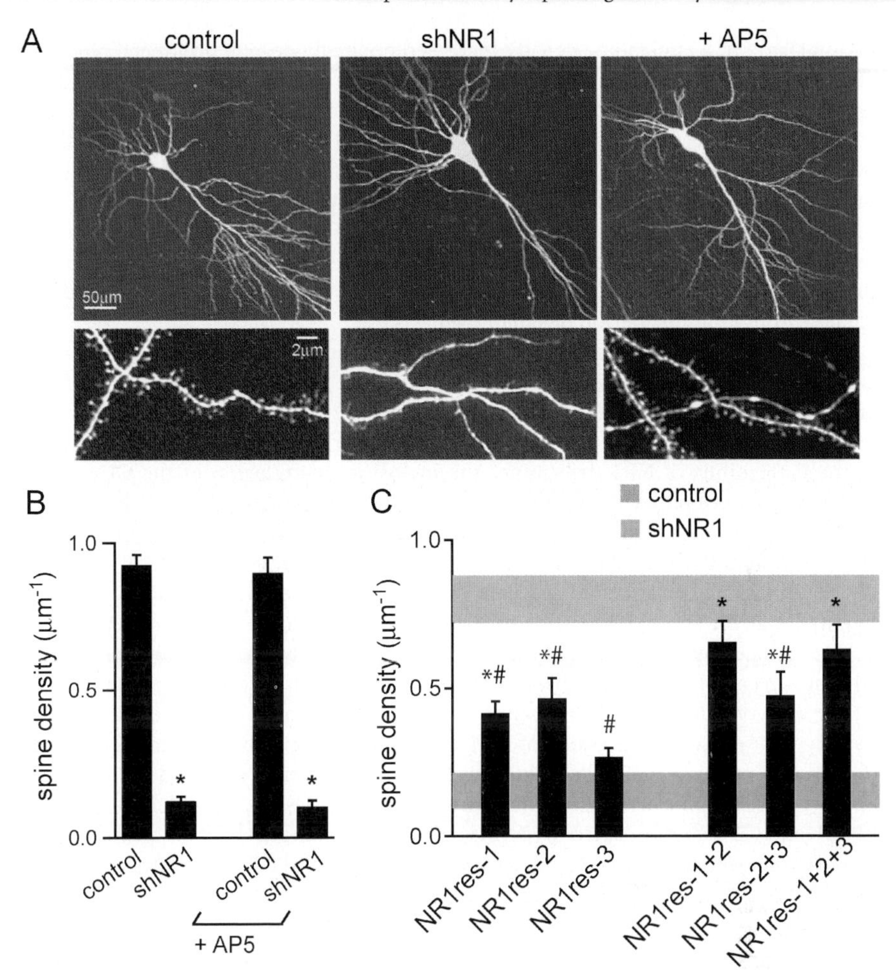

Fig. 5. Loss of dendritic spines after elimination of synaptic NMDARs is not mimicked by receptor antagonists and is prevented by the expression of specific NR1 splice isoforms. **A.** Top row, hippocampal pyramidal neurons in organotypic slices imaged with 2PLSM at 20 d post transfection with GFP (left), shNR1 (middle), and GFP-transfected neurons incubated with NMDARs antagonist AP5 for 20 d (right). *Scale bar* = 50 µm. Bottom row, higher magnification images of dendrites from GFP control, shNR1 and control + AP5. *Scale bar* = 2 µm. **B.** Summary of spine densities for GFP control and shNR1 neurons when incubated with and without AP5 for 20 d. $*$, $P < 0.05$ compared to GFP control neurons by ANOVA and Student-Newman-Keuls multiple comparisons test. **C.** Summary of spine densities for pyramidal neurons transfected with shNR1 and NR1 rescue splice isoforms 1, 2, or 3 at 15 DPT (*black bars*). $NR1_{res}$ denotes the NR1 rescue clones that carry silent mutations in the region targeted by shNR1. *Shaded horizontal bars* show the average $\pm 2*$SEM of spine density for control (*gray*) and shNR1 (*red*) neurons at 15 DPT. $*$ and #, $P < 0.05$ compared to shNR1 and control neurons, respectively, by ANOVA and Student-Newman-Keuls multiple comparisons test

nation (Fig. 5A, B) and spine density increased normally despite prolonged blockade of the receptor. This was the case whether the AP5 treatment was started at 10 DPT and continued for 10 days (spine density = $0.92 \pm 0.03\,\mu m^{-1}$ in untreated controls at 20 DPT and $0.90 \pm 0.05\,\mu m^{-1}$ in 10 day AP5/20 DPT neurons; $n = 10-12$ cells) or when AP5 incubation was started on the day of transfection (3 DIV) and maintained until imaging (spine density = $0.80 \pm 0.04\,\mu m^{-1}$ untreated controls at 15 DPT and $0.76 \pm 0.1\,\mu m^{-1}$ in 15 day AP5/15DPT neurons; $n = 8-9$ cells). The failure of the receptor antagonist treatment to induce spine loss demonstrates that the ionotropic properties of the NMDARs are not required to maintain normal density of dendritic spines throughout this period of synaptogenesis.

Native, functional NMDARs are heteromeric assemblies composed of two NR1 subunits and two NR2 subunits (Monyer et al. 1992; Laube et al. 1998; Schorge and Colquhoun 2003). The NR1 gene generates multiple splice isoforms that confer unique properties to the receptor assembly (Laurie and Seeburg 1994). Splicing of the C-terminus of NR1 determines the inclusion or exclusion of the C1 cassette and expression of either the C2 or C2' cassette, which regulates NMDAR trafficking, surface expression, signaling to the nucleus, and PDZ-based interactions (Ehlers et al. 1995,1998; Standley et al. 2000; Wenthold et al. 2003; Bradley et al. 2006).

To understand whether NR1 splice variants with different C-terminus tails differentially affect spine density, we tested the ability of different isoforms to rescue shNR1-induced spine loss (Fig. 5C). Expression of NR1-1_{res}, which contains the C0, C1, and C2 cassettes, increased spine density ~2-fold but did not restore spine density to wild-type levels ($0.41 \pm 0.04\,\mu m^{-1}$, $n = 9$ cells). Similar partial rescue of spine density was seen with expression of NR1-2_{res}, which contains the C0 and C2 cassettes ($0.47 \pm 0.07\,\mu m^{-1}$, $n = 12$ cells). Thus the presence or absence of C1 did not affect the ability to rescue spine density. Furthermore, expression of NR1-3_{res}, which contains the C0, C1, and C2' cassettes, failed to increase spine density above shNR1-expressing neurons ($0.27 \pm 0.03\,\mu m^{-1}$, $n = 10$; Fig. 5C). Since NR1-1_{res} and NR1-3_{res} differ only in the inclusion of the C2 or C2' cassette, these results suggest that expression of NR1 subunits with the C2 cassette, and not the C2' cassette, is necessary for spine maintenance.

To examine if multiple NR1 isoforms cooperatively promote spine growth, the rescue efficiency of co-expressing combinations of NR1 isoforms was tested. Co-expression of splice variants 1 and 2 returned spine density to near wild-type levels ($0.65 \pm 0.07\,\mu m^{-1}$, $n = 17$ cells; Fig. 5C), suggesting a cooperative effect on spine maintenance. Interestingly, the splice variants 1 and 2 are the predominant forms of NR1 subunits in hippocampal pyramidal neurons (Laurie and Seeburg 1994; Paupard et al. 1997). Conversely, co-expression of splice variant 3 with 2 alone or with 1 and 2 together did not increase spine density beyond that seen without the inclusion of variant 3. These experiments indicate that the assembly of isoforms 1 and 2 of the NR1 subunit is required for stable formation of dendritic spines and that this process is neither enhanced nor inhibited by the expression of isoform 3.

To conclude, we can say that NMDARs regulate neuronal function at many levels and through multiple pathways. First, opening of NMDARs contributes to postsynaptic depolarization and shapes circuit firing. Second, Ca influx through open receptors locally regulates synaptic properties and globally regulates gene transcription and cell survival. Third, NMDAR-dependent protein–protein interactions play a role in the assembly of the postsynaptic density (PSD). Here, we show that non-ionotropic

NMDAR-dependent signaling through specific C-tail splice isoforms regulates the long-term stability and density of dendritic spines as well as the numbers of excitatory synapses.

References

Alvarez VA, Sabatini BL (2007) Anatomical and hysiological lasticity of dendritic spines. Annu Rev Neurosci 30:79–97

Alvarez VA, Ridenour DA, Sabatini BL (2007) Distinct structural and ionotrophic roles of NMDA receptors in controlling spine and synapse stability. J Neurosci 27(28): 7365–76.

Bloodgood BL, Sabatini BL (2007a) Nonlinear regulation of unitary synaptic signals by CaV(2.3) voltage-sensitive calcium channels located in dendritic spines. Neuron 53:249–260.

Bloodgood BL, Sabatini BL (2007b) Ca(2+) signaling in dendritic spines. Curr Opin Neurobiol 17:345–351

Bradley J, Carter SR, Rao VR, Wang J, Finkbeiner S (2006) Splice variants of the NR1 subunit differentially induce NMDA receptor-dependent gene expression. J Neurosci 26:1065–1076.

Carter AG, Sabatini BL (2004) State-dependent calcium signaling in dendritic spines of striatal medium spiny neurons. Neuron 44:483–493.

Cummings JA, Mulkey RM, Nicoll RA, Malenka RC (1996) Ca2+ signaling requirements for long-term depression in the hippocampus. Neuron 16:825–833.

Ehlers MD, Tingley WG, Huganir RL (1995) Regulated subcellular distribution of the NR1 subunit of the NMDA receptor. Science 269:1734–1737.

Ehlers MD, Fung ET, O'Brien RJ, Huganir RL (1998) Splice variant-specific interaction of the NMDA receptor subunit NR1 with neuronal intermediate filaments. J Neurosci 18:720–730.

Engert F, Bonhoeffer T (1999) Dendritic spine changes associated with hippocampal long-term synaptic plasticity. Nature 399:66–70.

Fox K, Schlaggar BL, Glazewski S, O'Leary DD (1996) Glutamate receptor blockade at cortical synapses disrupts development of thalamocortical and columnar organization in somatosensory cortex. Proc Natl Acad Sci USA 93:5584–5589.

Harris KM (1999) Structure, development, and plasticity of dendritic spines. Curr Opin Neurobiol 9:343–348.

Hsieh H, Boehm J, Sato C, Iwatsubo T, Tomita T, Sisodia S, Malinow R (2006) AMPAR removal underlies Abeta-induced synaptic depression and dendritic spine loss. Neuron 52:831–843.

Huang L, Pallas SL (2001) NMDA Antagonists in the Superior Colliculus Prevent Developmental Plasticity But Not Visual Transmission or Map Compression. J Neurophysiol 86:1179–1194.

Jourdain P, Fukunaga K, Muller D (2003) Calcium/calmodulin-dependent protein kinase II contributes to activity-dependent filopodia growth and spine formation. J Neurosci 23:10645–10649.

Kim KS, Miller DL, Sapienza VJ, Chen C-MJ, Bai C, Grundke-Iqbal I, Currie JR, Wisniewski HM (1988) Production and characterization of monoclonal antibodies reactive to synthetic cerebrovascular amyloid peptide. Neurosci Res Comm 2:1212–1130.

Knott GW, Quairiaux C, Genoud C, Welker E (2002) Formation of dendritic spines with GABAergic synapses induced by whisker stimulation in adult mice. Neuron 34:265–273.

Kornau HC, Schenker LT, Kennedy MB, Seeburg PH (1995) Domain interaction between NMDA receptor subunits and the postsynaptic density protein PSD-95. Science 269:1737–1740.

Kuo YM, Emmerling MR, Vigo-Pelfrey C, Kasunic TC, Kirkpatrick JB, Murdoch GH, Ball MJ, Roher AE (1996) Water-soluble Abeta (N-40, N-42) oligomers in normal and Alzheimer disease brains. J Biol Chem 271:4077–4081.

Laube B, Kuhse J, Betz H (1998) Evidence for a tetrameric structure of recombinant NMDA receptors. J Neurosci 18:2954–2961.

Laurie DJ, Seeburg PH (1994) Regional and developmental heterogeneity in splicing of the rat brain NMDAR1 mRNA. J Neurosci 14:3180–3194.
Malenka RC, Nicoll RA (1993) NMDA-receptor-dependent synaptic plasticity: multiple forms and mechanisms. Trends Neurosci 16:521–527.
Malenka RC, Bear MF (2004) LTP and LTD: an embarrassment of riches. Neuron 44:5–21.
Maletic-Savatic M, Malinow R, Svoboda K (1999) Rapid dendritic morphogenesis in CA1 hippocampal dendrites induced by synaptic activity. Science 283:1923–1927.
Matsuzaki M, Honkura N, Ellis-Davies GC, Kasai H (2004) Structural basis of long-term potentiation in single dendritic spines. Nature 429:761–766.
McIlhinney RA, Philipps E, Le Bourdelles B, Grimwood S, Wafford K, Sandhu S, Whiting P (2003) Assembly of N-methyl-D-aspartate (NMDA) receptors. Biochem Soc Trans 31:865–868.
McLaurin J, Golomb R, Jurewicz A, Antel JP, Fraser PE (2000) Inositol stereoisomers stabilize an oligomeric aggregate of Alzheimer amyloid beta peptide and inhibit abeta-induced toxicity. J Biol Chem 275:18495–18502.
McLean CA, Cherny RA, Fraser FW, Fuller SJ, Smith MJ, Beyreuther K, Bush AI, Masters CL (1999) Soluble pool of Abeta amyloid as a determinant of severity of neurodegeneration in Alzheimer's disease. Ann Neurol 46:860–866.
Monyer H, Sprengel R, Schoepfer R, Herb A, Higuchi M, Lomeli H, Burnashev N, Sakmann B, Seeburg PH (1992) Heteromeric NMDA receptors: molecular and functional distinction of subtypes. Science 256:1217–1221.
Moolman DL, Vitolo OV, Vonsattel JP, Shelanski ML (2004) Dendrite and dendritic spine alterations in Alzheimer models. J Neurocytol 33:377–387.
Mucke L, Masliah E, Yu GQ, Mallory M, Rockenstein EM, Tatsuno G, Hu K, Kholodenko D, Johnson-Wood K, McConlogue L (2000) High-level neuronal expression of abeta 1-42 in wild-type human amyloid protein precursor transgenic mice: synaptotoxicity without plaque formation. J Neurosci 20:4050–4058.
Nagerl UV, Eberhorn N, Cambridge SB, Bonhoeffer T (2004) Bidirectional activity-dependent morphological plasticity in hippocampal neurons. Neuron 44:759–767.
Naslund J, Haroutunian V, Mohs R, Davis KL, Davies P, Greengard P, Buxbaum JD (2000) Correlation between elevated levels of amyloid beta-peptide in the brain and cognitive decline. JAMA 283:1571–1577.
Ngo-Anh TJ, Bloodgood BL, Lin M, Sabatini BL, Maylie J, Adelman JP (2005) SK channels and NMDA receptors form a Ca2+-mediated feedback loop in dendritic spines. Nature Neurosci 8:642–649.
Paupard MC, Friedman LK, Zukin RS (1997) Developmental regulation and cell-specific expression of N-methyl-D-aspartate receptor splice variants in rat hippocampus. Neuroscience 79:399–409.
Podlisny MB, Ostaszewski BL, Squazzo SL, Koo EH, Rydell RE, Teplow DB, Selkoe DJ (1995) Aggregation of secreted amyloid beta-protein into sodium dodecyl sulfate-stable oligomers in cell culture. J Biol Chem 270:9564–9570.
Sabatini BL, Oertner TG, Svoboda K (2002) The life-cycle of Ca^{2+} ions in spines. Neuron 33:439–452
Sawtell NB, Frenkel MY, Philpot BD, Nakazawa K, Tonegawa S, Bear MF (2003) NMDA Receptor-Dependent Ocular Dominance Plasticity in Adult Visual Cortex. Neuron 38:977–985.
Schorge S, Colquhoun D (2003) Studies of NMDA receptor function and stoichiometry with truncated and tandem subunits. J Neurosci 23:1151–1158.
Selkoe DJ (2001) Alzheimer's disease: genes, proteins, and therapy. Physiol Rev 81:741–766.
Shankar GM, Bloodgood BL, Townsend M, Walsh DM, Selkoe DJ, Sabatini BL (2007) Natural oligomers of the Alzheimer amyloid-beta protein induce reversible synapse loss by modulating an NMDA-type glutamate receptor-dependent signaling pathway. J Neurosci 27:2866–2875.

Snyder EM, Nong Y, Almeida CG, Paul S, Moran T, Choi EY, Nairn AC, Salter MW, Lombroso PJ, Gouras GK, Greengard P (2005) Regulation of NMDA receptor trafficking by amyloid-beta. Nature Neurosci 8:1051–1058.

Spires TL, Grote HE, Garry S, Cordery PM, Van Dellen A, Blakemore C, Hannan AJ (2004) Dendritic spine pathology and deficits in experience-dependent dendritic plasticity in R6/1 Huntington's disease transgenic mice. Eur J Neurosci 19:2799–2807.

Standley S, Roche KW, McCallum J, Sans N, Wenthold RJ (2000) PDZ domain suppression of an ER retention signal in NMDA receptor NR1 splice variants. Neuron 28:887–898.

Tavazoie SF, Alvarez VA, Ridenour DA, Kwiatkowski DJ, Sabatini BL (2005) Regulation of neuronal morphology and function by the tumor suppressors Tsc1 and Tsc2. Nature Neurosci 8:1727–1734.

Toni N, Buchs PA, Nikonenko I, Bron CR, Muller D (1999) LTP promotes formation of multiple spine synapses between a single axon terminal and a dendrite. Nature 402:421–425.

Walsh DM, Tseng BP, Rydel RE, Podlisny MB, Selkoe DJ (2000) The oligomerization of amyloid beta-protein begins intracellularly in cells derived from human brain. Biochemistry 39:10831–10839.

Wenthold RJ, Prybylowski K, Standley S, Sans N, Petralia RS (2003) Trafficking of NMDA receptors. Annu Rev Pharmacol Toxicol 43:335–358.

Zhou Q, Homma KJ, Poo MM (2004) Shrinkage of dendritic spines associated with long-term depression of hippocampal synapses. Neuron 44:749–757.

Zuo Y, Yang G, Kwon E, Gan WB (2005) Long-term sensory deprivation prevents dendritic spine loss in primary somatosensory cortex. Nature 436:261–265.

Soluble Oligomers of the Amyloid β-Protein: Impair Synaptic Plasticity and Behavior

Dennis J. Selkoe[1]

Summary. The central quest of research on Alzheimer's disease (AD) is to identify precisely what molecular process first interferes with episodic declarative memory and then prevent that process. During the last 25 years, neuropathological, biochemical, genetic, cell biological and even therapeutic studies in humans have all supported the hypothesis that the gradual cerebral accumulation of soluble and insoluble assemblies of the amyloid β-protein (Aβ) in limbic and association cortices triggers a cascade of biochemical and cellular alterations that produce the clinical phenotype. Missense mutations in presenilin, an unusual intramembrane aspartyl protease, cause rare forms of early-onset AD by altering its cleavage of the amyloid precursor protein (APP) transmembrane domain to increase the ratio of Aβ42 to Aβ40 peptides, thereby promoting Aβ oligomerization. The reasons for elevated cortical Aβ42 levels in most patients with typical, late-onset AD are unknown, but based on recent work, these could turn out to include augmented neuronal release of Aβ during some kinds of synaptic activity. Elevated levels of soluble Aβ42 monomers enable formation of soluble oligomers that can diffuse into synaptic clefts. We have identified certain APP-expressing cultured cell lines that form low-n oligomers intracellularly and release a portion of them into the medium. We find that these naturally secreted soluble oligomers – at picomolar concentrations – can disrupt hippocampal LTP in slices and in vivo and can also impair the memory of a complex learned behavior in rats. Aβ trimers appear to be more potent in disrupting LTP than are dimers. The cell-derived oligomers also decrease dendritic spine density in organotypic hippocampal slice cultures, and this decrease can be prevented by administration of Aβ antibodies or small-molecule modulators of Aβ aggregation. The signaling pathways mediating these effects are under study. Intensive attempts to develop safe and effective Aβ-lowering agents have brought us into human trials that have provided preliminary evidence of cognitive benefit. This therapeutic progress has been accompanied by advances in imaging the Aβ deposits non-invasively in humans. A new diagnostic-therapeutic paradigm to successfully address AD and its harbinger, mild cognitive impairment-amnestic type, is emerging.

Introduction

During most of the 20th century, neurodegenerative diseases remained among the most enigmatic disorders of medicine. The scientific study of these conditions was descriptive in nature, detailing the clinical and neuropathological phenotypes associated with various diseases, but etiologies and pathogenic mechanisms remained obscure. Beginning in the 1970s, advances in two principal areas – biochemical pathology and molecular genetics – combined to yield powerful clues to the molecular underpinnings of several previously "idiopathic" brain disorders. Among the classical neurodegenerative diseases, perhaps the most rapid progress occurred in research on Alzheimer's

[1] Center for Neurologic Diseases, Department of Neurology, Brigham and Women's Hospital, Harvard Medical School, Boston, MA 0211,5 USA, email: dselkoe@rics.bwh.harvard.edu

Selkoe et al.
Synaptic Plasticity and the Mechanism of Alzheimer's Disease
© Springer-Verlag Berlin Heidelberg 2008

disease (AD). In disorders like Huntington's disease, amyotrophic lateral sclerosis and even Parkinson's disease, unbiased genetic screens, linkage analysis and positional cloning have identified causative genes that subsequently allowed the formulation of specific biochemical hypotheses. In sharp contrast, modern research on AD developed in the opposite order: the identification of the protein subunits of the classical brain lesions guided geneticists to disease-inducing genes, for example, APP, apolipoprotein E and tau. Thus, a biochemical hypothesis of disease – that AD is a progressive cerebral amyloidosis caused by the aggregation of the amyloid β-protein (Aβ) – preceded and enabled the discovery of etiologies.

As progress in deciphering genotype-to-phenotype relationships in AD accelerated during the last two decades, it became apparent that the key challenge for understanding and ultimately treating AD was to focus not on what was killing neurons over the course of the disease but rather on what was interfering subtly and intermittently with episodic declarative memory well before widespread neurodegeneration had occurred (Selkoe 2002). In other words, one wishes to understand the factors underlying early synaptic dysfunction in the hippocampus and then attempt to neutralize these as soon as feasible, perhaps even before a definitive diagnosis of AD can be made. This steady movement of the field toward ever-earlier stages of the disorder is exemplified by the recognition and intensive study of minimal cognitive impairment–amnestic type (MCI; Petersen et al. 1999). And yet patients who die with a diagnosis of MCI have been found to already have a histopathology essentially indistinguishable from classical AD (Price and Morris 1999). Therefore, even earlier phases of this continuum are likely to become recognized, and these might show milder histopathology and might have biochemically, but not yet microscopically, detectable Aβ species that mediate synaptic dysfunction.

The IPSEN symposium for which this volume serves as a record focused on bringing together investigators at the forefront of elucidating the structure and function of hippocampal synapses with investigators focused on understanding how early assemblies of Aβ may compromise some of these synapses. This chapter will summarize some of the observations and discoveries made by the author and his colleagues over several years that have the goal of identifying the earliest synaptotoxic molecules in Alzheimer's disease – and neutralizing them.

Moving from Synthetic Aβ Peptides to Naturally Secreted Aβ Assemblies

A wealth of data from many laboratories now supports the once controversial hypothesis that the accumulation and aggregation of Aβ initiates a complex cascade of molecular and cellular changes that gradually leads to the clinical features of MCI-amnestic type and then frank Alzheimer's disease (Hardy and Selkoe 2002; Hardy and Higgins 1992; Selkoe 1991). As a result, understanding precisely how Aβ accumulation and assembly compromise synaptic structure and function has become the centerpiece of therapeutically oriented research on the disease.

A great many studies have been conducted using synthetic Aβ peptides of either 40 or 42 amino acids, mimicking the two most common lengths of Aβ found in normal human brain and in the cortical and vascular amyloid deposits of AD patients. But naturally generated Aβ peptides in brain, cerebrospinal fluid (CSF) or the media of cultured cells are considerably more heterogeneous in length (see, e.g., Busciglio et al. 1993; Haass et al. 1992; Iwatsubo et al. 1994; Podlisny et al. 1995; Saido et al. 1995; Seubert et al. 1992; Shoji et al. 1992; Vigo-Pelfrey et al. 1993; Wang et al. 1996). More importantly, almost all studies of synthetic Aβ p peptides have employed concentrations upwards of 1 uM, often 10–40 uM, because the critical concentration allowing relatively rapid assembly of synthetic Aβ40 (the most commonly used peptide) into amyloid fibrils is in the high nanomolar range or greater. Also, synthetic Aβ peptides are often solublized, at least initially, in highly non-physiological solvents (acetonitrile, trifluoroacetic acid, sodium hydroxide, etc.), since these allow the hydrophobic, relatively insoluble peptide to be dissolved and used experimentally.

Many different types of assembly forms of synthetic Aβ, including amyloid fibrils, protofibrils (PF), annular structures, paranuclei, Aβ-derived diffusible ligands and globulomers, have been described over the last two decades (for reviews, see Caughey and Lansbury 2003; Teplow 1998). Even some of the smaller synthetic aggregates do not fulfill the definition of soluble oligomers: Aβ assemblies that are not pelleted from physiological fluids by high-speed centrifugation. For example, protofibrils are intermediates that were observed in the course of studying the fibrillization of synthetic Aβ (Harper et al. 1997; Hartley et al. 1999; Walsh et al. 1999). They are flexible structures that can continue to polymerize in vitro to form amyloid fibrils or can de-polymerize to lower-order species. Protofibrils are narrower than bona fide amyloid fibrils (~4–5 nm versus 8–10 nm). Ultrastructural analyses of synthetic protofibril preparat by electron microscopy (EM) and atomic force microscopy (AFM) have revealed both straight and curved assemblies up to 150 nm in length. Synthetic Aβ protofibrils have been shown to contain substantial β-sheet structure, as they are able to bind Congo red or Thioflavin T in ordered fashion.

Annular assemblies of synthetic Aβ are donut-like structures with an outer diameter of 8–12 nm and an inner diameter of 2.0–2.5nm that are distinct from protofibrils by AFM and EM (Bitan and Teplow 2005; Lashuel et al. 2002). Smaller oligomeric species of synthetic Aβ than protofibrils and annuli have been observed, depending on how synthetic Aβ is prepared and incubated, and some have been designated Aβ-derived diffusible ligands (ADDLs; Lambert et al. 1998). Apparent ADDL-like oligomeric assemblies have been isolated from postmortem AD brains and their presence correlated with memory loss (Gong et al. 2003). In separate work, chemical stabilization of synthetic Aβ42 assembly intermediates has revealed an apparent hexamer periodicity, with hexamer, dodecamer, and octadecamer structures observed (Bitan et al. 2003).

In striking contrast to the properties of these various synthetic assemblies, Aβ peptides that are generated in vivo by humans and lower mammals or by cultured cells are diverse with regard to their N- and C-termini, occur naturally in extracellular fluids at low to sub-nanomolar concentrations (Gravina et al. 1995; Näslund et al. 2000; Scheuner et al. 1996; Walsh et al. 2002) and can begin to assemble into metastable dimers, trimers and higher oligomers while still at low nanomolar levels (Podlisny et al. 1995; Walsh et al. 2000). Dimeric, trimeric and apparently tetrameric soluble oligomers have been described in cultured cells (Podlisny et al. 1995; Walsh et al. 2000), and SDS-stable oligomers of varying sizes have also been detected by Western blotting in APP transgenic mouse brain and human brain (Enya et al. 1999; Funato et al. 1999; Kawarabayashi et al. 2004; Lesne et al. 2006; McLean et al. 1999; Roher et al. 1996). Such natural (i.e., non-synthetic) Aβ oligomers can be resistant not only to SDS but also to chaotropic salts like guanidine hydrochloride and to the Aβ-degrading protease IDE (insulin-degrading enzyme), which can only efficiently digest monomeric Aβ (Walsh et al. 2002). Aβ oligomers produced by cultured cells could be related to the recently described Aβ*56 species, which represents a soluble, SDS-stable dodecamer found in the brains of at least some APP transgenic mouse lines (Lesne et al. 2006). Like the Aβ oligomers produced from cultured cells (Walsh et al. 2002), Aβ∗56 can disrupt synaptic function and thus affect memory (Lesne et al. 2006). Whether Aβ∗56, the cell-derived dimers and trimers (Podlisny et al. 1995) and other soluble oligomers observed in biological systems represent stable assemblies of solely Aβ under native conditions or whether such small oligomeric assemblies stably associate with one or more "carrier" proteins in vivo is currently unclear. Upon further study, Aβ∗56 and Aβ trimers secreted by cultured cells could turn out to share common synaptotoxic properties.

Naturally Secreted Aβ Oligomers Abrogate Hippocampal Synaptic Plasticity

In collaboration with the laboratory of Michael Rowan, we have taken advantage of our discovery that Chinese hamster ovary (CHO) cells stably expressing the AD-causing Val717Phe mutation in APP secrete soluble oligomers detectable on SDS gels as dimers, trimers and tetramers (Podlisny et al. 1995) to conduct a series of studies defining the electrophysiological activities of these assemblies. As mentioned above, we view these cell-derived, low-n oligomers as having several advantages over synthetic Aβ aggregates, most notably their natural production by these and other cells at low- to sub-nanomolar concentrations. Importantly, the CHO-derived oligomers are entirely soluble: centrifuging the conditioned medium at >100 000 g for > hour still leaves all of the monomers and oligomers in the supernatant, with retained neurobiological activity. We have confirmed the identity of these species as bona fide Aβ oligomers using radiosequencing, immunoprecipitation by both N- and C-terminal-specific Aβ monoclonal antibodies, and non-denaturing size exclusion chromatography (SEC). The oligomers can be readily detect in CHO cells stably expressing wild-type human APP (Podlisny et al. 1995), and their disease relevance is supported by our finding that expressing either an APP mutation, a presenilin 1 mutation or a presenilin 2 mutation in CHO cells elevates Aβ42 levels in the media and increases the levels of the SDS-stable dimers and trimers (Xia et al. 1997). We are currently attempting to determine the atomic masses of the oligomers by mass spectrometry to establish their bonding structure; we have already confirmed the mass of the secreted 4 kDa species in the medium as human Aβ monomer, but larger quantities are being prepared to try to obtain the masses of the dimers and trimers.

Dominic Walsh (now at University College Dublin), who led many of our studies characterizing the biochemical and neurobiological properties of the cell-derived oligomers while he was a fellow in my laboratory, established our very productive collaboration with Michael Rowan and his fellow, Igor Kluybin, at Trinity College Dublin. We conducted experiments together to determine whether the monomers and oligomers present in the conditioned medium of the APP V717F-expressing CHO cells (called 7PA2 cells) can interfere with basal synaptic transmission and/or long-term potentiation (LTP) in the hippocampus in vivo. Through a cannula implanted in the lateral ventricle of adult rats, we administered tiny amounts (usually 1.5–5.0 ul) of the straight conditioned medium collected from the 7PA2 cells. Basal synaptic transmission was unaffected by the medium, but CA1 field recordings after high-frequency stimulation (HFS) of the CA3 Schaeffer collateral-to-CA1 pathway revealed a highly consistent failure to maintain an LTP response in the presence of the 7PA2 medium but not with the medium of untransfected CHO cells (called CHO- cells; Walsh et al. 2002). In the presence of the 7PA2 medium, synaptic potentiation would initially be induced by the HFS but waned over the next 1–3 hours, so that LTP was not maintained. We next performed a number of experiments that indicated that the soluble Aβ monomers in the 7PA2 medium were not responsible for the failure to maintain LTP but the soluble oligomers were. For example, we took advantage of the ability of IDE to degrade the monomers but leave the oligomers (which are surprisingly stable) behind, and this preparation still inhibited hippocampal LTP in vivo. Conversely, injecting the medium after treating the 7PA2 cells with a γ-secretase inhibitor at low doses that reduced the monomers by just ~40% but thereby decreased dimer/trimer levels by ~90% led to a normal LTP response. We concluded that secreted, low-n oligomers of human Aβ at picomolar concentrations and in the absence of monomers as well as higher order aggregates (protofibrils, fibrils) could potently inhibit the maintenance of LTP in the mammalian hippocampus.

In recent work, our laboratory has extended our electrophysiological findings to hippocampal slices in vitro. We confirmed the effect of the 7PA2 medium in inhibiting LTP in slices from wild-type mice (Townsend et al. 2006). Further, the cell-derived oligomers blocked a chemically induced form of LTP caused by stimulating dissociated mouse hippocampal neurons with picrotoxin and glycine in the absence of added magnesium (Townsend et al. 2007). The use of the chemically induced LTP paradigm (chem-LTP) on neuronal cultures allowed us to measure the biochemical effects of the soluble oligomers on activation of several different signaling kinases. We found that of five kinases we evaluated, three (ERK MAPK CaMKII and Akt/PKB) had clearly reduced activation upon chem-LTP, whereas two others (PKA and PKC) were unaffected. These data begin to reveal some of the signaling pathways that can be interrupted by naturally secreted oligomers, indicating that there is specificity to the hippocampal response and suggesting that there are discrete receptor-mediated pathways through which soluble oligomers act.

We have also developed protocols for separating the secreted oligomers by SEC under non-denaturing conditions (Klyubin et al. 2005; Townsend et al. 2006; Walsh et al. 2005). When as many as 100 SEC fractions of the 7PA2 conditioned medium were obtained and then tested separately on mouse hippocampal slices, we observed that fractions rich in trimers inhibited LTP even more potently than did fractions enriched principally in dimers; agina, monomers were without effect (Townsend et al. 2006). These results suggest again the existence of a level of molecular specificity in the interactions of Aβ oligomers with neuronal targets. With Dominic Walsh and his colleagues, we are now attempting to purifiy the oligomers from the CHO cell medium to homogeneity so that they might be labeled and used in ligand binding studies (e.g., by autoradiography) to try to identify the cognate receptors for oligomers but not monomers.

Cell-derived Oligomers Interfere with the Memory of a Complex Learned Behavior

In collaboration with James Cleary and Karen Ashe at the University of Minnesota, we have been able to demonstrate significant cognitive deficits in a complex lever pressing task in adult rats that are directly attributable to a naturally secreted assembly form of Aβ (Cleary et al. 2005). The active Aβ species were the soluble oligomers, not monomers and in the absence of protofibrils and fibrils, and the oligomer effects were characterized by rapid onset, high potency and transience. These combined biochemical and cognitive analyses provided the first direct behavioral data supporting the emerging hypothesis that diffusible oligomers of Aβ are responsible for important components of neuronal dysfunction leading up to or associated with AD (reviewed in Klein et al. 2001; Selkoe 2002; Walsh et al. 2003). The rapid onset and transience of the soluble oligomer effects we saw in the rats are indicative of a pathophysiological action, similar to that seen with amnestic drugs such as scopolamine. Such pathophysiological activities can occur independently of structural neuronal injury and, indeed, our experimental paradigm does not suggest a role for any neurodegenerative processes in the deleterious cognitive effects produced transiently by low picomolar quantities of soluble Aβ oligomers.

In studies reported prior to the Cleary et al. (2005) paper, intracerebral injections of synthetic Aβ that included mixtures of Aβ fibrils, protofibrils, oligomers and monomers in indeterminate proportions exerted deleterious effects on learned behavior in rats (Cleary et al. 1995; Frautschy et al. 2001; O'Hare et al. 1999). However, the deficits were reported to develop over weeks and, once present, they persisted or worsened even without further injections. Because intracerebral injections are usually made during stereotaxic surgery (rather than long after such surgery, as in our paradigm), any immediate acute effects of the injectate on learning or memory typically go unstudied. Two studies attempted to address potential acute effects of very high concentrations of synthetic Aβ species. In one study, increased maze entry errors were observed after i.c.v. injections of 2.0 l of 1 mM synthetic Aβ(1-40) that were given immediately before maze testing (Sweeney et al. 1997). In another study, no effect on intrahippocampal injections of 0.5 μl of 1 mM synthetic Aβ(1-40) that were administered 30 minutes before testing were observed in a radial arm maze test (McDonald et al. 1994). The concentrations and total amounts of synthetic Aβ administered were several orders of magnitude higher than those of the natural, cell-secreted oligomers used in our study (Cleary et al. 2005). Also, the duration between administration and testing was different. Because the assembly sizes of the synthetic Aβ that was injected were not defined in either study, the specific forms of Aβ being tested could not be deduced, in contrast to the present work.

Our work may provide an explanation for the observation that subtle brain dysfunction can be detected in certain individuals who are genetically at risk for AD but remain neurologically stable for many years before the expected onset of disease (Bookheimer et al. 2000; Small et al. 2000). In addition, our linkage of acute behavioral deficits with soluble oligomers provides a mechanistic explanation for the rapid restoration of cognitive function that followed administration of anti-Aβ antibodies in behaviorally impaired APP transgenic mice (Dodart et al. 2002; Kotilinek et al. 2002).

Cell-Derived Oligomers Decrease Dendritic Spine Density in Hippocampus by an NMDA-Dependent Signaling Pathway

Our next approach to deciphering the synaptic effects of natural Aβ oligomers was to ask whether they can induce structural alterations of synapses in association with the clear functional deficits described in the previous two sections above. We exposed organotypic rat hippocampal slice cultures that had been biolistically transfected with EGFP to subnanomolar concentrations of SEC-separated 7PA2 cell Aβ monomers or dimers/trimers for periods varying from 1 to 15 days. We observed a marked decrease in dendritic spine density in EGFP-positive pyramidal neurons that reached ~60% loss after 15 days of exposure (Shankar et al. 2007). Electrophysiological experiments confirmed that the decreased spine density reflected a loss of excitatory synapses. Miniature EPSCs were measured from neurons in hippocampal slices treated for 10–15 days. mEPSC amplitude, which reflects the postsynaptic AMPA-type glutamate receptor (AMPAR)-mediated response to the release of a single vesicle of glutamate, was slightly but significantly reduced in oligomer-treated neurons compared to control-medium treated neurons. Moreover, as expected from the loss of dendritic spines, the inter-mEPSC interval was significantly increased (195 ± 5 ms vs. 85 ± 3 ms), corresponding to a ~50% reduction in mEPSC frequency (Shankar et al. 2007).

The strong decrease in dendritic spine density was reversible when the Aβ oligomer-rich medium (which had been applied for 10 days) was exchanged for normal slice medium for 5 days; spine density returned to almost normal levels (Shankar et al. 2007). Spine loss was also fully prevented by co-application to the slices of an N-terminal monoclonal antibody to Aβ (6E10) together with the soluble oligomers; denatured (boiled) 6E10 had no rescue effect. We also examined *scyllo*-inositol (AZD-103), a *myo*-inositol steroisomer shown by Joanne McLaurin and colleagues to decrease Aβ plaque number and size and prevent behavioral deficits in an APP transgenic mouse line (McLaurin et al. 2006). At 5 uM concentrations, *scyllo*-inositol co-administered with the soluble oligomers fully prevented the spine loss, whereas a known inactive stereoisomer, *chiro*-inositol, did nothing.

Acute application of synthetic Aβ can trigger a rapid reduction of NMDAR-dependent currents in dissociated cortical neurons that is thought to result from Aβ-mediated activation of nicotinic acetylcholine receptors (nAchRs; Snyder et al. 2005). To determine if the same signaling pathway was responsible for the Aβ oligomer-mediated spine loss, we examined the ability of NMDAR and nAchR antagonists to mimic or block the effects of secreted Aβ. Chronic blockade of nAchRs by 10-day exposure to the irreversible antagonist α-Bungarotoxin had no effect on the spine density of control neurons and did not affect the degree of spine loss produced by oligomer application. Conversely, although 10 days of application of the NMDAR antagonist CPP (20 μM) alone had no effect on spine density, it completely prevented the spine loss normally seen with Aβ oligomer incubation, indicating that NMDAR-activity is required for Aβ-mediated spine loss (Shankar et al. 2007).

To determine if Aβ oligomers reduce NMDAR-mediated synaptic Ca influx in hippocampal pyramidal neurons, 2-photon laser photoactivation (2PLP) of caged (MNI)-glutamate was used to stimulate the postsynaptic terminal on a visualized spine while evoked calcium transients were measured in the spine head with 2P laser scanning microscopy (Bloodgood and Sabatini 2007). Whole-cell recordings were obtained from

CA1 pyramidal neurons in acute hippocampal slices, and cells were filled through the patch pipette with the Ca-sensitive, green-fluorescing fluorophore Fluo-5F and the Ca-independent, red-fluorescing fluorophore Alexa Fluor-594. Red fluorescence was used to visualize morphology and to select spines within the proximal 150 μm of an apical dendrite for analysis. Uncaging-evoked fluorescence transients were monitored in the spine head and adjacent dendrite while uncaging-evoked postsynaptic potentials (uEPSPs) were recorded at the soma. Fluorescence transients were quantified relative to maximal green fluorescence at saturating levels of Ca ($\Delta G_{uEPSP}/G_{sat}$), a measure that, under our recording conditions, is linearly proportional to evoked changes in Ca ($\Delta[\mathrm{Ca}]_{uEPSP}$; Bloodgood and Sabatini 2007). We found that $\Delta[\mathrm{Ca}]_{uEPSP}$ was smaller in the presence of Aβ oligomers ($\Delta G_{uEPSP}/G_{sat} = 7.4 \pm 0.5\%$, 21/5 spines/cells) than in control conditions ($\Delta G_{uEPSP}/G_{sat} = 10.2 \pm 1.1\%$, 20/7 spines/cells, $p < 0.05$). In contrast, $\Delta[\mathrm{Ca}]_{uEPSP}$ in the presence of Aβ monomers ($\Delta G_{uEPSP}/G_{sat} = 9.9 \pm 1.0\%$, 19 spines/6 cells) was the same as in control conditions. Thus, exposure to picomolar levels of soluble Aβ oligomers acutely reduces but does not abolish $\Delta[\mathrm{Ca}]_{uEPSP}$, consistent with a partial reduction of NMDAR-mediated Ca influx into active spines.

Partial blockade of NMDARs and reduced Ca^{2+} influx through NMDARs favor the induction of long-term depression (LTD) via a calcineurin-dependent pathway (Cummings et al. 1996; Mulkey et al. 1994)). LTD induction is accompanied by a shrinkage of dendritic spines that is mediated by regulation of the actin-depolymerization factor cofilin (Shankar et al. 2007; Zhou et al. 2004) at a conserved serine (position 3) that can be phosphorylated by LIM-kinase. We found that neurons transfected with a plasmid encoding cofilin S3D, in which serine 3 is replaced by the phosphomimetic amino acid aspartate, rendering cofilin constitutively inactive, displayed a net increase in dendritic spine density. Expression of this construct also prevented the loss of dendritic spines normally seen following a 10-day exposure to Aβ oligomers. To examine whether Aβ oligomers mediate spine loss through activation of calcineurin, we examined the ability of FK506 to block Aβ oligomer-induced spine loss. Incubation in FK506 (1 uM) alone for 24 hr had no effect on spine density. However, co-administration of FK506 with Aβ oligomers fully prevented spine loss (Shankar et al. 2007).

Conclusions

Through a series of systematic studies of soluble oligomers of human Aβ secreted by cultured cells, we have documented that low-n oligomers – but not monomers from the same source and at higher concentrations – can inhibit LTP without affecting basal synaptic transmission, can reversibly alter the structure of excitatory synapses by decreasing spines, and can interfere with the memory of a learned behavior in healthy adult rats. We interpret these data to signify that small diffusible oligomers impair both synaptic function and synaptic structure on glutamatergic neurons in the hippocampus. Therapeutically, we show that oligomer-mediated spine loss is prevented by antibodies to Aβ and a small molecule inhibitor of Aβ aggregation. We had previously reported that both active and passive immunotherapy of normal adult rats protect these animals form the LTP-inhibiting effects of soluble Aβ oligomers administered intracerebroventricularly (Klyubin et al. 2005). Mechanistically, we find that dendritic spine reduction requires activity of a signaling cascade involving NMDARs, calcineurin, and cofilin. These results suggest that Aβ oligomers shift the activation of NMDAR-dependent signaling cascades towards pathways involved in the induction of LTD. We propose that chronic activation of these pathways by soluble Aβ oligomers found in the brains of patients with AD contributes to the pathogenesis of the disease and may underlie the synapse loss in hippocampus that occurs early in the disease process and is a key quantitative correlate of degree of cognitive impairment in AD patients (Davies et al. 1987; Masliah et al. 2001; Terry et al. 1991).

Overall, our studies with soluble natural oligomers suggest a model (Fig. 1) in which exposure to Aβ oligomers mimics a state of partial NMDAR blockade, either by reducing NMDAR activation, reducing NMDAR-dependent Ca influx, or enhancing NMDAR-dependent activation of calcineurin. We propose that these effects promote the LTD-inducing mode and inhibit the LTP-inducing mode of NMDAR-dependent signaling. Since LTP induction promotes spine enlargement and the growth of new spines whereas LTD induction promotes spine shrinkage and retraction (Engert and Bonhoeffer 1999; Maletic-Savatic et al. 1999; Matsuzaki et al. 2004; Nagerl et al. 2004; Zhou et al. 2004), this oligomer-induced imbalance would promote the progressive loss of dendritic spines and glutamatergic synapses. Furthermore, a previous study reported that FK506 prevented the ability of synthetic Aβ to inhibit LTP in hippocampus (Chen et al. 2002), suggesting that the inhibition of LTP by Aβ oligomers reflects a shift in the LTP/LTD balance rather than a direct blockade of LTP-inducing pathways.

Our proposed model is largely consistent with a recent report that also concludes that Aβ decreases dendritic spine density and alters glutamatergic signaling in CA1 pyramidal neurons in rat hippocampal organotypic slices (Hsieh et al. 2006). Hsieh et al. found that neurons overexpressing human APP have decreased dendritic spine density compared to control neurons. Furthermore, transduction with a C-terminal fragment of APP (β-CTF) decreased AMPAR-mediated synaptic currents. This effect mimicked and partially occluded metabotropic glutamate receptor-induced LTD, involving endocytosis of GluR2-containing AMPARs. Important differences between the two studies may explain why we found no decrease in AMPAR-mediated mEPSCs whereas Hsieh et al. found a ~30% reduction in AMPAR EPSCs. For example, Hsieh et al. performed recordings ~24 hours after transduction with β-CTF and therefore described effects of short-term exposure to Aβ that may be mechanistically different from the longer-term effects described here. Furthermore, Hsieh et al. found alterations in synaptic transmission in neurons expressing β-CTF but not in neighboring cells, suggesting cell-autonomous or autocrine effects of APP or Aβ. Aβ in their system is generated intracellularly and secreted, so at least some of the observed effects could be attributable to Aβ acting intracellularly, whereas our system involves extracellular application of secreted Aβ. The APP transduction paradigm used by Hsieh et al. does not allow a distinction between the effects of Aβ oligomers and monomers, whereas we are able to ascribe the synaptic changes specifically to soluble oligomers of Aβ. However, despite the differences in the design and analysis performed in the two studies, both support a model in which Aβ perturbs excitatory synapses by enhancing LTD in an activity- and calcineurin-dependent manner.

The next step in this line of research is to translate this approach and the resultant findings to the human disease. We must determine whether soluble Aβ monomers and oligomers isolated directly from the cerebral cortex of typical, late-onset AD cases can induce the same structural and functional deficits and whether they do so by

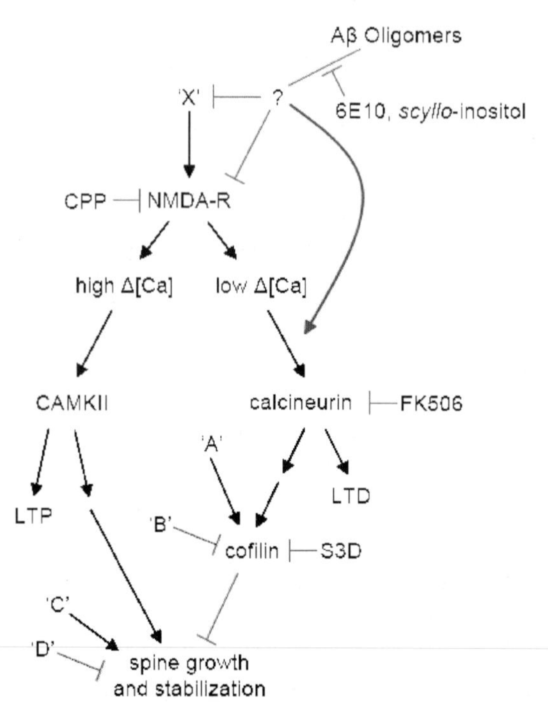

Fig. 1. Proposed pathways that regulate spine density and that are affected by Aβ oligomers, based on the results of our results. Ca^{2+} influx through synaptic NMDARs can activate at least two pathways that regulate spine density. On the left-hand side, high levels of Ca^{2+} accumulation, such as those reached during tetanic or suprathreshold synaptic stimulation, induce LTP via a CAMKII-dependent pathway (reviewed in Nicoll and Malenka 1999). LTP-inducing stimuli also trigger the enlargement of dendritic spines and growth of new spines in a NMDAR- and CAMKII-dependent manner (Engert and Bonhoeffer 1999; Jourdain et al. 2003; Maletic-Savatic et al. 1999; Matsuzaki et al. 2004; Nagerl et al. 2004). Introduction of active CAMKII in neurons is sufficient to induce new spine growth (Jourdain et al. 2003). In the right-hand side pathway, low levels of Ca^{2+} accumulation, such as those reached during low-frequency subthreshold stimulation, induce LTD through a calcineurin-dependent pathway (reviewed in Malenka and Bear 2004). LTD-inducing stimuli also lead to spine shrinkage via an NMDAR/calcineurin/cofilin-dependent pathway and spine retraction through an NMDAR-dependent pathway (Nagerl et al. 2004; Zhou et al. 2004). The calcineurin and cofilin dependence of LTD-associated spine retraction have not been examined. In this model, full block of NMDARs interrupts both pathways, leading to no net spine loss. Partial block of NMDARs favors activation of the right-hand pathway, LTD induction, and loss of spines. In addition, multiple factors (A, B, C, and D) act independently of NMDARs, CAMKII, and calcineurin to regulate cofilin and spine density. We find that soluble Aβ oligomers decrease spine density in an NMDAR/calcineurin/cofilin-dependent manner, consistent with activation of the pathway shown on the right. Aβ oligomers reduce NMDAR-dependent Ca^{2+} transients, possibly shifting stimuli that normally activate the left-hand pathway to instead activate those on the right. This activation might occur through direct interaction of Aβ with NMDARs or by first activating unknown factors (X) that may lead to inhibition of NMDAR-mediated synaptic Ca^{2+} influx. Aβ may also facilitate NMDAR-dependent activation of calcineurin via additional pathways. *Blue lines* indicate levels at which soluble Aβ oligomers may modulate the pathway and *red lines* indicate elements of the pathway tested in this study

similar signaling mechanisms as summarized above. If this can be achieved, we may be able to conclude that we have identified the "smoking gun" that induces synaptic impairment in the brains of Alzheimer's patients. In a sense, this next step would be attempting to fulfil one of Koch's postulates for identifying the causative agent of an idiopathic disease: isolating the putative agent, transmitting it to normal brain tissue (of a rodent, as this cannot be done in humans), and showing that the agent reproduces the key phenotypic features of the disorder. If such experiments are successful, they will strongly validate and encourage the current intensive effort to treat, and perhaps ultimately prevent, AD using agents that lower Aβ in the brain.

References

Bitan G, Kirkitadze MD, Lomakin A, Vollers SS, Benedek GB, Teplow DB (2003) Amyloid β–protein (Aβ) assembly: Aβ 40 and Aβ 42 oligomerize through distinct pathways. Proc Natl Acad Sci USA 100:330–335.

Bitan G, Teplow DB (2005) Preparation of aggregate-free, low molecular weight amyloid-beta for assembly and toxicity assays. Methods Mol Biol 299:3–9.

Bloodgood BL, Sabatini BL (2007) Nonlinear regulation of unitary synaptic signals by CaV(2.3) voltage-sensitive calcium channels located in dendritic spines. Neuron 53:249–260.

Bookheimer SY, Strojwas MH, Cohen MS, Saunders AM, Pericak-Vance MA, Mazziotta JC, Small GW (2000) Patterns of brain activation in people at risk for Alzheimer's disease. N Engl J Med 343:450–456.

Busciglio J, Gabuzda DH, Matsudaira P, Yankner BA (1993) Generation of β-amyloid in the secretory pathway in neuronal and nonneuronal cells. Proc Natl Acad Sci. USA 90:2092–2096.

Caughey B, Lansbur, PT (2003) Protofibrils, pores, fibrils, and neurodegeneration: separating the responsible protein aggregates from the innocent bystanders. Annu Rev Neurosci 26:267–298.

Chen F, Gu Y, Hasegawa H, Ruan X, Arawaka S, Fraser P, Westaway D, Mount H, St George-Hyslop P (2002). Presenilin 1 mutations activate gamma 42-secretase but reciprocally inhibit epsilon-secretase cleavage of amyloid precursor protein (APP) and S3-cleavage of notch. Biol Chem 277:36521–36526.

Cleary J, Hittner JM, Semotuk M, Mantyh P, O'Hare E (1995) Beta-amyloid(1-40) effects on behavior and memory. Brain Res 682:69–74.

Cleary JP, Walsh D, Hofmeister JJ, Shankar GM, Kuskowski MA, Selkoe DJ, Ashe KH (2005) Natural oligomers of the amyloid β-protein specifically disrupt cognitive function. Nature Neurosci 8:79–84.

Cummings BJ, Pike CJ, Shankle R, Cotman CW (1996). Beta-amyloid deposition and other measures of neuropathology predict cognitive status in Alzheimer's disease. Neurobiol Aging 17:921–933.

Davies CA, Mann DM, Sumpter PQ, Yates PO (1987). A quantitative morphometric analysis of the neuronal and synaptic content of the frontal and temporal cortex in patients with Alzheimer's disease. J Neurol Sci 78:151–164.

Dodart JC, Bales KR, Gannon KS, Greene SJ, DeMattos RB, Mathis C, DeLong CA, Wu S, Wu X, Holtzman DM, Paul SM (2002) Immunization reverses memory deficits without reducing brain Abeta burden in Alzheimer's disease model. Nature Neurosci 5:452–457.

Engert F, Bonhoeffer T (1999) Dendritic spine changes associated with hippocampal long-term synaptic plasticity. Nature 399:66–70.

Enya M, Morishima-Kawashima M, Yoshimura M, Shinkai Y, Kusui K, Khan K, Games D, Schenk D, Sugihara S, Yamaguchi H, Ihara Y (1999) Appearance of sodium dodecyl sulfate-stable amyloid beta-protein (Abeta) dimer in the cortex during aging. Am J Pathol 154:271–279.

Frautschy SA, Hu W, Kim P, Miller SA, Chu T, Harris-White ME, Cole GM (2001) Phenolic anti-inflammatory antioxidant reversal of Abeta-induced cognitive deficits and neuropathology. Neurobiol Aging 22:993–1005.

Funato H, Enya M, Yoshimura M, Morishima-Kawashima M, Ihara Y (1999) Presence of sodium dodecyl sulfate-stable amyloid beta-protein dimers in the hippocampus CA1 not exhibiting neurofibrillary tangle formation. Am J Pathol 155:23–28.

Gong Y, Chang L, Viola KL, Lacor PN, Lambert MP, Finch CE, Krafft GA, Klein WL (2003) Alzheimer's disease-affected brain: presence of oligomeric A beta ligands (ADDLs) suggests a molecular basis for reversible memory loss. Proc Natl Acad Sci USA 100:10417–10422.

Gravina SA, Ho L, Eckman CB, Long KE, Otvos LJ, Younkin LH, Suzuki N, Younkin SG (1995). Amyloid β protein (Aβ) in Alzheimer's disease brain. J Biol Chem 270:7013–7016.

Haass C, Schlossmacher M, Hung AY, Vigo-Pelfrey C, Mellon A, Ostaszewski B, Lieberburg I, Koo EH, Schenk D, Teplow D, Selkoe D (1992) Amyloid β-peptide is produced by cultured cells during normal metabolism. Nature 359:322–325.

Hardy J, Selkoe, DJ (2002). The amyloid hypothesis of Alzheimer's disease: progress and problems on the road to therapeutics. Science 297:353–356.

Hardy JA, Higgins GA (1992). Alzheimer's disease: the amyloid cascade hypothesis. Science 256:184–185.

Harper JD, Wong SS, Lieber CM, Lansbury Jr PT (1997) Observation of metastable Aβ amyloid protofibrils by atomic force microscopy. Chem Biol 4:119–125.

Hartley D, Walsh DM, Ye CP, Diehl T, Vasquez S, Vassilev PM, Teplow DB, Selkoe DJ (1999) Protofibrillar intermediates of amyloid β-protein induce acute electrophysiological changes and progressive neurotoxicity in cortical neurons. J Neurosci 19:8876–8884.

Hsieh H, Boehm J, Sato C, Iwatsubo T, Tomita T, Sisodia S, Malinow R (2006) AMPAR removal underlies Abeta-induced synaptic depression and dendritic spine loss. Neuron 52:831–843.

Iwatsubo T, Odaka A, Suzuki N, Mizusawa H, Nukina H, Ihara Y (1994) Visualization of A beta 42(43) and A beta 40 in senile plaques with end-specific A beta monoclonals: evidence that an initially deposited species is A beta 42(43). Neuron 13:45–53.

Jourdain P, Fukunaga K, Muller D (2003) Calcium/calmodulin-dependent protein kinase II contributes to activity-dependent filopodia growth and spine formation. J Neurosci 23:10645–10649.

Kawarabayashi T, Shoji M, Younkin LH, Wen-Lang L, Dickson DW, Murakami T, Matsubara E, Abe K, Ashe KH, Younkin SG (2004) Dimeric amyloid beta protein rapidly accumulates in lipid rafts followed by apolipoprotein E and phosphorylated tau accumulation in the Tg2576 mouse model of Alzheimer's disease. J Neurosci 24:3801–3809.

Klein WL, Krafft GA, Finch CE (2001) Targeting small Abeta oligomers: the solution to an Alzheimer's disease conundrum? Trends Neurosci 24:219–224.

Klyubin I, Walsh DM, Lemere CA, Cullen WK, Shankar GM, Betts V, Spooner ET, Jiang L, Anwyl R, Selkoe DJ, Rowan MJ (2005) Amyloid beta protein immunotherapy neutralizes Abeta oligomers that disrupt synaptic plasticity in vivo. Nature Med 11:556–561.

Kotilinek LA, Bacskai B, Westerman M, Kawarabayashi T, Younkin L, Hyman BT, Younkin S, Ashe KH (2002) Reversible memory loss in a mouse transgenic model of Alzheimer's disease. J Neurosci 22:6331–6335.

Lambert MP, Barlow AK, Chromy BA, Edwards C, Freed R, Iosatos M, Morgan TE, Rozovsky I, Trommer B, Viola KL, Wals P, Zhang C, Finch CE, Krafft GA, Klein WL (1998) Diffusible, nonfribrillar ligands derived from $A\beta_{1-42}$ are potent central nervous system neurotoxins. Proc Natl Acad Sci USA 95:6448–6453.

Lashuel HA, Hartley D, Petre BM, Walz T, Lansbury PT Jr. (2002) Neurodegenerative disease: amyloid pores from pathogenic mutations. Nature 418:291.

Lesne S, Koh MT, Kotilinek L, Kayed R, Glabe CG, Yang A, Gallagher M, Ashe KH (2006) A specific amyloid-beta protein assembly in the brain impairs memory. Nature 440:352–357.

Malenka RC, Bear MF (2004) LTP and LTD: an embarrassment of riches. Neuron 44:5–21.

Maletic-Savatic M, Malinow R, Svoboda K (1999) Rapid dendritic morphogenesis in CA1 hippocampal dendrites induced by synaptic activity. Science 283:1923–1927.

Masliah E, Mallory M, Alford M, DeTeresa R, Hansen LA, McKeel DW Jr., Morris JC (2001) Altered expression of synaptic proteins occurs early during progression of Alzheimer's disease. Neurology 56:127–129.

Matsuzaki M, Honkura N, Ellis-Davies GC, Kasai H (2004). Structural basis of long-term potentiation in single dendritic spines. Nature 429:761–766.

McDonald MP, Dahl EE, Overmier JB, Mantyh P, Cleary J (1994) Effects of an exogenous beta-amyloid peptide on retention for spatial learning. Behav Neural Biol 62:60–67.

McLaurin J, Kierstead ME, Brown ME, Hawkes CA, Lambermon MH, Phinney AL, Darabie AA, Cousins JE, French JE, Lan MF, Chen F, Wong SS, Mount HT, Fraser PE, Westaway D, St George-Hyslop P (2006) Cyclohexanehexol inhibitors of Abeta aggregation prevent and reverse Alzheimer phenotype in a mouse model. Nature Med 12:801–808.

McLean CA, Cherny RA, Fraser FW, Fuller SJ, Smith MJ, Beyreuther K, Bush AI, Masters CL (1999) Soluble pool of Abeta amyloid as a determinant of severity of neurodegeneration in Alzheimer's disease. Ann Neurol 46:860–866.

Mulkey RM, Endo S, Shenolikar S, Malenka RC (1994) Involvement of a calcineurin/inhibitor-1 phosphatase cascade in hippocampal long-term depression. Nature 369:486–488.

Nagerl UV, Eberhorn N, Cambridge SB, Bonhoeffer T (2004) Bidirectional activity-dependent morphological plasticity in hippocampal neurons. Neuron 44:759–767.

Näslund J, Haroutunian V, Mohs R, Davis K, Davies P, Greengard P, Buxbaum J (2000) Correlation between elevated levels of amyloid β-peptides in the brain and cognitive decline. JAMA 283:1571–1577.

Nicoll RA, Malenka RC (1999) Expression mechanisms underlying NMDA receptor-dependent long-term potentiation. Ann NY Acad Sci 868:515–525.

O'Hare E, Weldon DT, Mantyh PW, Ghilardi JR, Finke MP, Kuskowski MA, Maggio JE, Shephard RA, Cleary J (1999) Delayed behavioral effects following intrahippocampal injection of aggregated A beta (1-42). Brain Res 815:1–10.

Petersen RC, Smith GE, Waring SC, Ivnik RJ, Tangalos EG, Kokmen E (1999) Mild cognitive impairment: clinical characterization and outcome. Arch Neurol 56:303–308.

Podlisny MB, Ostaszewski BL, Squazzo SL, Koo EH, Rydel RE, Teplow DB, Selkoe DJ (1995) Aggregation of secreted amyloid β-protein into SDS-stable oligomers in cell culture. J Biol Chem 270:9564–9570.

Price JL, Morris JC (1999) Tangles and plaques in nondemented aging and "preclinical" Alzheimer's disease. Ann Neurol 45:358–368.

Roher AE, Chaney MO, Kuo Y-M, Webster SD, Stine WB, Haverkamp LJ, Woods AS, Cotter RJ, Tuohy JM, Krafft GA, Bonnell BS, Emmerling MR (1996) Morphology and toxicity of Aβ-(1-42) dimer derived from neuritic and vascular amyloid deposits of Alzheimer's disease. J Biol Chem 271:20631–20635.

Saido TC, Iwatsubo T, Mann DMA, Shimada H, Ihara Y, Kawashima S (1995) Dominant and differential deposition of distinct β-amyloid peptide species, $A\beta_{N3(p3)}$, in senile plaques. Neuron 14:457–466.

Scheuner D, Eckman C, Jensen M, Song X, Citron M, Suzuki N, Bird TD, Hardy J, Hutton M, Kukull W, Larson E, Levy-Lahad E, Viitanen M, Peskind E, Poorkaj P, Schellenberg G, Tanzi R, Wasco W, Lannfelt L, Selkoe D, Younkin S (1996).Secreted amyloid β-protein similar to that in the senile plaques of Alzheimer's disease is increased *in vivo* by the presenilin 1 and 2 and APP mutations linked to familial Alzheimer's disease. Nature Med 2:864–870.

Selkoe DJ (1991) The molecular pathology of Alzheimer's disease. Neuron 6:487–498.

Selkoe DJ (2002) Alzheimer's disease is a synaptic failure. Science 298:789–791.

Seubert P, Vigo-Pelfrey C, Esch F, Lee M, Dovey H, Davis D, Sinha S, Schlossmache, MG, Whaley J, Swindlehurst C, McCormack R, Wolfert R, Selkoe DJ, Lieberburg I, Schenk D (1992) Isolation and quantitation of soluble Alzheimer's β-peptide from biological fluids. Nature 359:325–327.

Shankar GM, Bloodgood BL, Townsend M, Walsh DM, Selkoe DJ, Sabatini BL (2007) Natural oligomers of the Alzheimer amyloid-beta protein induce reversible synapse loss by modulating an NMDA-type glutamate receptor-dependent signaling pathway. J Neurosci 27:2866–2875.
Shoji M, Golde TE, Ghiso J, Cheung TT, Estus S, Shaffer LM, Cai X, McKay DM, Tintner R, Frangione B, Younkin SG (1992) Production of the Alzheimer amyloid β protein by normal proteolytic processing. Science 258:126–129.
Small GW, Ercoli LM, Silverman DH, Huang SC, Komo S, Bookheimer SY, Lavretsky H, Miller K, Siddarth P, Rasgon NL, Mazziotta JC, Saxena S, Wu HM, Mega MS, Cummings JL, Saunders AM, Pericak-Vance MA, Roses AD, Barrio JR, Phelps ME (2000) Cerebral metabolic and cognitive decline in persons at genetic risk for Alzheimer's disease. Proc Natl Acad Sci USA 97:6037–6042.
Snyder EM, Nong Y, Almeida CG, Paul S, Moran T, Choi EY, Nairn AC, Salter MW, Lombroso PJ, Gouras GK, Greengard P (2005) Regulation of NMDA receptor trafficking by amyloid-beta. Nature Neurosci 8:1051–1058.
Sweeney WA, Luedtke J, McDonald MP, Overmier JB (1997) Intrahippocampal injections of exogenous beta-amyloid induce postdelay errors in an eight-arm radial maze. Neurobiol Learn Mem 68:97–101.
Teplow DB (1998) Structural and kinetic features of amyloid beta-protein fibrillogenesis. Amyloid 5:121–142.
Terry RD, Masliah E, Salmon DP, Butters N, DeTeresa R, Hill R, Hansen LA, Katzma, R (1991) Physical basis of cognitive alterations in Alzheimer's disease: synapse loss is the major correlate of cognitive impairment. Ann Neurol 30:572–580.
Townsend M, Shankar GM, Mehta T, Walsh DM, Selkoe DJ (2006) Effects of secreted oligomers of amyloid beta-protein on hippocampal synaptic plasticity: a potent role for trimers. J Physiol 572:477–492.
Townsend M, Mehta T, Selkoe, DJ (2007). Soluble amyloid β-protein inhibits specific signal transduction cascades common to the insulin reception pathway. J Biol Chem, in press.
Vigo-Pelfrey C, Lee D, Keim PS, Lieberburg I, Schenk D (1993) Characterization of β-amyloid peptide from human cerebrospinal fluid. J Neurochem 61:1965–1968.
Walsh D, Klyubin I, Fadeeva J, Cullen WK, Anwyl R, Wolfe M, Rowan M, Selkoe D (2002) Naturally secreted oligomers of the Alzheimer amyloid β-protein potently inhibit hippocampal long-term potentiation *in vivo*. Nature 416:535–539.
Walsh DM, Hartley DM, Kusumoto Y, Fezoui Y, Condron MM, Lomakin A, Benedek GB, Selkoe DJ, Teplow DB (1999) Amyloid beta-protein fibrillogenesis. Structure and biological activity of protofibrillar intermediates. J Biol Chem 274:25945–25952.
Walsh DM, Tseng BP, Rydel RE, Podlisny MB, Selkoe DJ (2000) Detection of intracellular oligomers of amyloid β-protein in cells derived from human brain. Biochemistry 39:10831–10839.
Walsh DM, Hartley DM, Selkoe DJ (2003) The many faces of Aβ: Structures and activity. Curr Med Chem - Immun Endoc Metab Agents 3:277–291.
Walsh DM, Townsend TM, Podlisny MB, Shankar, GM, Fadeeva J, El-Agnaf O, Hartley DM, Selkoe DJ (2005) Certain inhibitors of synthetic Aβ fibrillogenesis block oligomerization of natural Aβ and thereby rescue long term potentiation. J Neurosci 25:2455–2462.
Wang R, Sweeney D, Gandy SE, Sisodia, SS (1996) The profile of soluble amyloid b protein in cultured cell media. Detection and quantificaiton of amyloid β protein and variants by immunoprecipitation-mass spectrometry. J Biol Chem 271:31894–31902.
Xia W, Zhang J, Kholodenko D, Citron M, Podlisny MB, Teplow DB, Haass C, Seubert P, Koo EH, Selkoe DJ (1997) Enhanced production and oligomerization of the 42-residue amyloid β-protein by Chinese hamster ovary cells stably expressing mutant presenilins. J Biol Chem 272:7977–7982.
Zhou Q, Homma KJ, Poo MM (2004). Shrinkage of dendritic spines associated with long-term depression of hippocampal synapses. Neuron 44:749–757.

Why Alzheimer's is a Disease of Memory: Synaptic Targeting by Pathogenic Aβ Oligomers (ADDLs)

William L. Klein[1], *Fernanda De Felice*[1], *Pascale N. Lacor*[1], *Mary P. Lambert*[1], *and Wei-Qin Zhao*[1,2]

Summary. Early Alzheimer's disease manifests as a crippling inability to form new memories, but why Alzheimer's is specific for memory has yet to be answered. As evidenced by this meeting, research increasingly focuses on deterioration of synapses and dendritic spines (1), a concept introduced more than 30 years ago by Scheibel and colleagues (2). This synaptic damage is now attributed to the impact of soluble Abeta oligomers, thanks to contributions from multiple laboratories. Abeta oligomers (here referred to as "ADDLs") were identified in 1998 as a new type of toxin, structurally distinct from amyloid fibrils, that rapidly prevented LTP (3). ADDL-induced disruption of plasticity is measurable at multiple levels, including ectopic over-expression of Arc, a spine cytoskeletal protein essential for memory formation. Confirming predictions based on the Arc response, ADDLs cause critical receptors to be eliminated from synaptic membranes and induce aberrations in spine morphology, with sustained presence of ADDLs resulting in spine elimination (4).

The mechanism underlying these synaptic pathologies likely holds the key to understanding why AD is specific for memory. ADDL-induced pathologies are not broad consequences of whole-scale neuronal deterioration but instead derive from a highly specific attachment to the spines of certain excitatory synapses. Whether obtained from AD brain or prepared in vitro, ADDLs bind to their targeted spines with high affinity, essentially acting as gain-of-function pathogenic ligands. Brain-derived and synthetic ligands are structurally equivalent 12mers that are strikingly elevated in AD brain (5) and also appear in animal models of AD, roughly concomitant with memory failure.

Recent investigations into the synaptic targets of ADDLs implicate NMDA receptors, insulin receptors, and neighboring synaptic proteins. Memantine, an NMDA receptor antagonist used as an AD therapeutic drug, effectively inhibits ADDL-induced pathologies in the short term, while antibodies against the NR1 subunit significantly reduce ADDL binding. These results suggest that ADDLs bind at or near NMDA receptors. Insulin receptors are implicated by findings that prior exposure of neurons to insulin results in virtually complete inhibition of ADDL binding. While unoccupied insulin receptors are essential for ADDL binding, they are not sufficient, implying that high affinity binding depends upon additional co-receptors, possibly comprising NR1 subunits or other nearby proteins. A synaptic response that may prove especially important for cognitive failure is a rapid and massive removal of insulin receptors from dendritic plasma membranes triggered when ADDLs are added prior to exogenous insulin. Accompanying this removal is an increase of insulin receptors within neuronal cell bodies, a net receptor redistribution that renders neurons insulin resistant. Antagonistic interaction between ADDLs and insulin provides a basis for possible CNS insulin resistance in AD and predicts therapeutic benefits for drugs that promote brain insulin function. Overall, knowing how ADDLs target and disrupt specific synapses will bring us closer to understanding why AD is a disease of memory and provide new avenues for the discovery of disease-modifying therapeutic drugs.

[1] Department of Neurobiology and Physiology, Cognitive Neurology and Alzheimer's Disease Center, Northwestern University, Evanston, IL 60208, email: wklein@northwestern.edu

[2] Alzheimer's Department, Merck Research Laboratories, Merck & Co. Inc., 770 Sumneytown Pike, PO Box4, 44K, West Point, PA 19486

Selkoe et al.
Synaptic Plasticity and the Mechanism of Alzheimer's Disease
© Springer-Verlag Berlin Heidelberg 2008

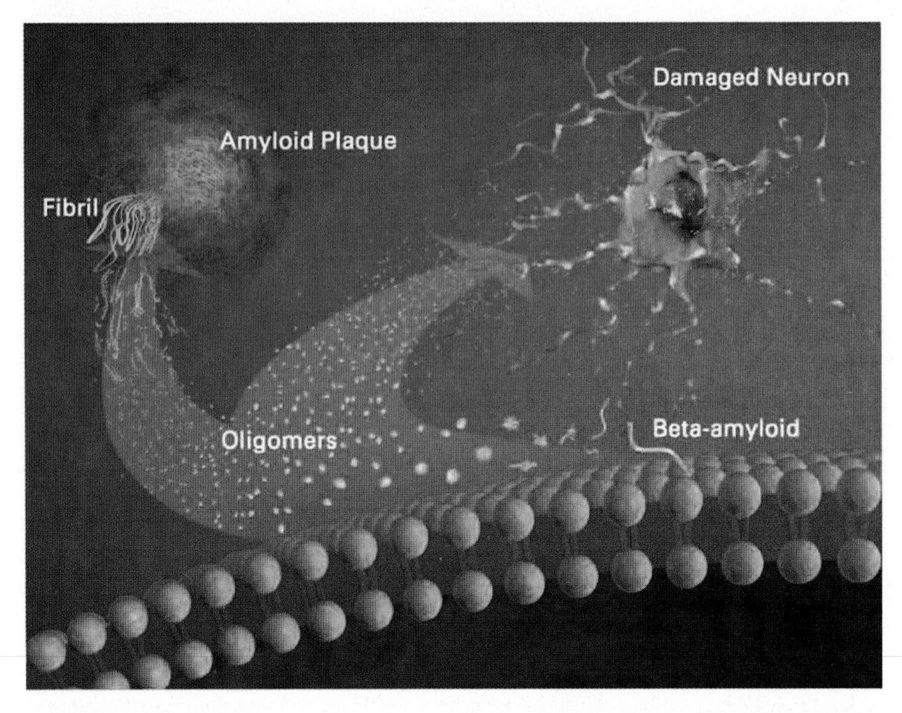

Fig. 1. Oligomers in pathway to harm. From HHS "Progress Report on Alzheimer's Disease" (Rodgers 2005)

Early Alzheimer's disease (AD) manifests as a crippling inability to form new memories (Selkoe 2002). Why memory in particular is targeted has long been a fundamental mystery of the disease mechanism, but new insights are beginning to emerge from investigations of novel, highly specific neurotoxins generated by the Aβ peptide. For many years, it had been thought that AD was initiated by the neurotoxicity of large insoluble Aβ fibrils found in amyloid plaques. New findings, however, make clear that fibrils are not the only toxic structures generated by self-association of Aβ and are probably not the most important ones. It now is increasingly likely that memory loss in AD can be attributed to diffusible Aβ oligomers (Fig. 1), which act in effect like pathogenic hormones. This chapter reviews the discovery that Aβ oligomers constitute a new type of toxin found in AD brain, and it describes the action of Aβ oligomers as gain-of-function pathogenic ligands capable of targeting particular synapses with great specificity and disrupting synapse shape, composition and the mechanisms of plasticity. It reviews findings that Aβ oligomers also initiate mechanisms leading to synapse loss, tau hyperphosphorylation, oxidative stress, and ultimately nerve cell death. It finishes with very recent findings that Aβ oligomers dramatically alter neuronal insulin receptor function, providing a basis for CNS insulin resistance in AD. By explaining why AD is specifically a disease of memory as well as accounting for major features of AD neuropathology, the toxicity of Aβ oligomers provides a unifying mechanism for AD pathogenesis and an optimum target for diagnostics and therapeutics.

Discovery of Toxic Oligomers and Their Relevance to Memory Loss in AD

Alternative outcomes of Aβ self-assembly – Fibrils and globular oligomers (ADDLs). The dominant theory for AD for more than 20 years has been the amyloid cascade hypothesis (Glenner and Wong 1984; Lewis et al. 1988; Masters and Beyreuther 1988; Hardy and Higgins 1992). Focusing on fibrillar Aβ this powerful hypothesis in its original form essentially comprised two fundamental tenets: 1) AD dementia was caused by neuronal death, and 2) neuronal death was triggered by cascades initiated by the large, insoluble amyloid fibrils. Consistent with this theory, synthetic Aβ fibrils structurally mimicking those found in disease pathology (Pike et al. 1991; Hilbich et al. 1991; Burdick et al. 1992) were found to kill neurons, whereas Aβ monomers did not (Pike et al. 1991, 1993; Lambert et al. 1994; Lorenzo and Yankner 1994). The important principle was established that Aβ toxicity was a gain-of-function pathology requiring self-assembly.

Despite its intuitive appeal and strong experimental support, the original amyloid cascade hypothesis suffered from significant inconsistencies (Terry 1994; Katzman et al. 1988; Klein et al. 2001). Two salient flaws are epitomized in studies of mouse AD models used for tests of passive vaccines. These mice are models of early AD, developing plaques and memory dysfunction as they age. Remarkably, when injected with antibodies against Aβ, the mice show a significant reversal of memory dysfunction. Recovery is evident in as little as 24 hours. Equally remarkable, the recovering mice exhibit no decrease in plaque burden (Dodart et al. 2002; Kotilinek et al. 2002). The memory-restoring effects of Aβ antibodies are difficult to reconcile with the tenets of the original cascade; the rapid recovery points to a mechanism not involving neuron death and the null impact on plaque indicates that the relevant Aβ toxins are species other than fibrils. In harmony with the conclusions from animal models, a lack of direct correlation between plaque burden and dementia in humans has long been noted by neuropathologists (Katzman et al. 1988), who have considered this lack to be a major flaw in the original amyloid cascade hypothesis, and the possibility that neuropil damage might be at least as important as nerve cell death was recognized by Scheibel and colleagues as early as 1975 (Scheibel et al. 1975).

Discovery of neurotoxic Aβ oligomers (ADDLs). Is there a simple mechanism to explain Aβ-dependent memory loss that is independent of neuron death and not instigated by the presence of amyloid plaques? An explanation lies in the earlier discovery of toxic Aβ oligomers. Oligomers are long-lived molecular entities, not simply intermediates in fibrillogenesis as originally thought, and they exhibit their own set of neurotoxic properties, including an impact on memory-related mechanisms of synaptic plasticity (Lambert et al. 1998).

In a seminal observation that ultimately led to the discovery of toxic oligomers, Finch and colleagues reported that formation of large Aβ aggregates could be blocked by apoJ (clusterin) without eliminating toxicity toward PC12 cells (Oda et al. 1995). Our subsequent investigation of Aβ-apoJ solutions determined that the relevant moieties are small globular Aβ oligomers and that these oligomers are potent CNS neurotoxins (Lambert et al. 1998). Alternative methods for generating non-fibrillar toxins sans apoJ have verified that the active species solely comprise assemblies of Aβ (Klein et al. 2001; Lambert et al. 2001; Walsh et al. 2002; Stine et al. 2003). Importantly, even after 24 hours

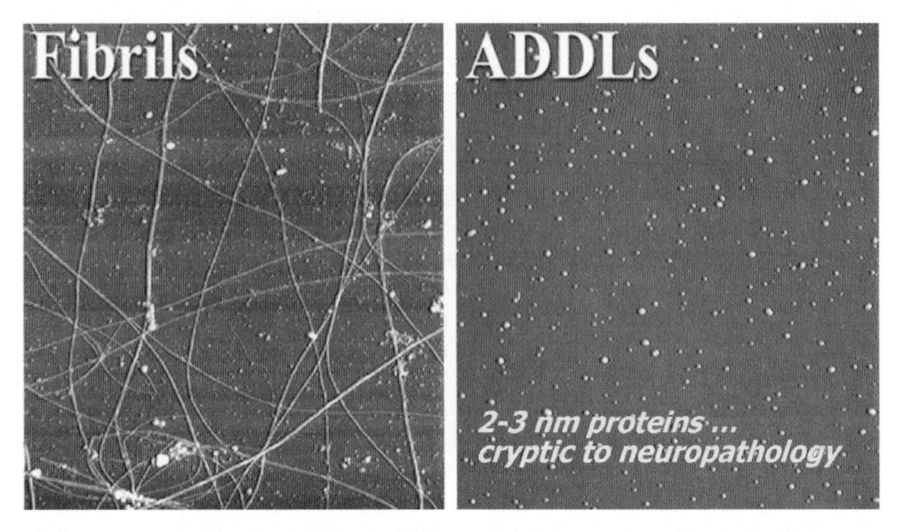

Fig. 2. Structural differences in Aβ preparations shown by atomic force microscopy. Depending on preparation method, assembled forms obtained from synthetic $A\beta_{1-42}$ provide either fibrillar preparations (A) or globular oligomers (B) (Klein et al. 2001)

at 37 degrees, toxic solutions examined by atomic force microscopy (AFM) exhibit no rod-like protofibrillar or fibrillar structures (Fig. 2). The toxic globular molecules have been named Abeta-derived diffusible ligands (ADDLs) to underscore that they represent a new type of toxin structurally and mechanistically distinct from fibrils and protofibrils. The association of toxicity with sub-fibrillar forms of disease-causing fibrillogenic proteins is now known to be a widely occurring phenomenon (Glabe 2006; Ferreira et al. 2007; Haass and Selkoe 2007), with ADDLs being only the first example.

The composition of Aβ oligomers formed in vitro comprises a spectrum of SDS-resistant oligomers. As detected by Western blots, these range from dimers to 24 mers, with the patterns showing marked antibody dependence. The commonly used 4G8 monoclonal, for example, is unusually sensitive to dimers whereas it poorly recognizes mid- to large-sized oligomers (Chromy et al. 2003; Lambert et al. 2001). Although it has been suggested that SDS-PAGE itself might generate small oligomers, pure monomers are readily detected using fresh preparations (Chromy et al. 2003). Not all oligomer states are equally abundant, and oligomer assembly itself is extremely sensitive to in vitro conditions, presumably reflecting the conformational dynamics of Aβ. By HPLC-SEC, oligomers can be separated into two fractions migrating with centers at ~15 kDa and ~50–60 kDa (Chromy et al. 2003). Both fractions, however, appear similar by SDS-PAGE, which yields predominantly monomers, 3mers and 4mers. Oligomers in the larger pool grow SDS stable when incubated at 37 degrees (Klein 2002) or with prostaglandins or levuglandins (Boutaud et al. 2006). At present, many basic questions concerning the mechanism of oligomer formation and the relationship between oligomerization and fibrillogenesis are unanswered, although novel approaches and hypotheses are generating significant new insights (Bitan et al. 2001; Vollers et al. 2005; Maynard et al. 2005).

Discovery that ADDLs inhibit LTP. Metastable Aβ oligomers are structurally intriguing, but the reason they have garnered major attention is that they selectively damage mechanisms germane to memory loss. In our initial studies of ADDLs and synaptic plasticity, we found that short-term exposure of hippocampal slices to ADDLs rapidly inhibits long-term potentiation (LTP; Klein et al. 1997; Lambert et al. 1998), the classic paradigm for memory-related synaptic mechanisms (Bennett 2000). Although primarily studied ex vivo, preliminary experiments showed that ADDLs injected stereotaxically into living mice also cause rapid and profound LTP inhibition (Lambert et al. 1998; Klein 2000). The initial impact of oligomers on synaptic plasticity is rapid and non-degenerative, which led us to predict that if oligomers were responsible for memory loss in early AD through an impact on synaptic plasticity, then early memory loss should be reversible (Klein et al. 2001; Lambert et al. 1998). This prediction was verified in the impressive immunization studies previously mentioned with transgenic mouse AD models (Kotilinek et al. 2002; Dodart et al. 2002). The impact of ADDLs also is selective for particular aspects of plasticity. For example, whereas oligomers inhibit LTP, they do not inhibit LTD (Wang et al. 2002). On the other hand, oligomers do block the reversal of LTD. The net neurological effect of oligomers in the hippocampus thus is to repress positive synaptic feedback. Inhibition of LTP formation and LTD reversal both would be expected to compromise memory formation (Klein et al. 2001).

A different preparation of oligomers, obtained from cultured media of hAPP-transfected CHO cells, has proven to be exceptionally potent at inhibiting LTP in vivo (Walsh et al. 2002). By Western blots, the cell-derived oligomers comprise mostly small, SDS-stable 3mers and 4mers. Whether antibodies sensitive to larger oligomers might detect 12mers or other species has not been reported. Controls using insulin-degrading enzyme, which proteolyzes monomers but not oligomers, ruled out the possibility that monomeric Aβ contributed to inhibition. Drugs that blocked Aβ production also blocked accumulation of the neurologically active molecules.

Most significantly, intracerebral microinjection of these naturally produced oligomers, at very low doses, produces memory dysfunction. This significant demonstration thus links oligomers, plasticity failure, and memory loss. Reflecting a major paradigm shift triggered by these and related studies, a new version of the amyloid cascade hypothesis has been proposed (Lambert et al. 1998; Klein et al. 2001; Hardy and Selkoe 2002). Its salient new features are that 1) early memory loss stems from synapse failure rather than neuron death and 2) synapse failure derives from the actions of soluble Aβ oligomers rather than fibrils.

Validation of Clinical Relevance

Conformation-dependent antibodies to detect ADDLs in AD. The findings reviewed so far have come from experimental models and oligomeric toxins generated in vitro. If soluble Aβ oligomers are to be valid targets for AD therapeutics, they must be present in human brain and manifest a strong AD-dependent accumulation. A critical question is whether AD-affected brains contain Aβ oligomers that are structurally and toxicologically identical to the neurologically active species studied ex vivo.

To address this question, our group introduced the use of conformation-dependent antibodies (Lambert et al. 2001). Rabbits injected with ADDL preparations produce

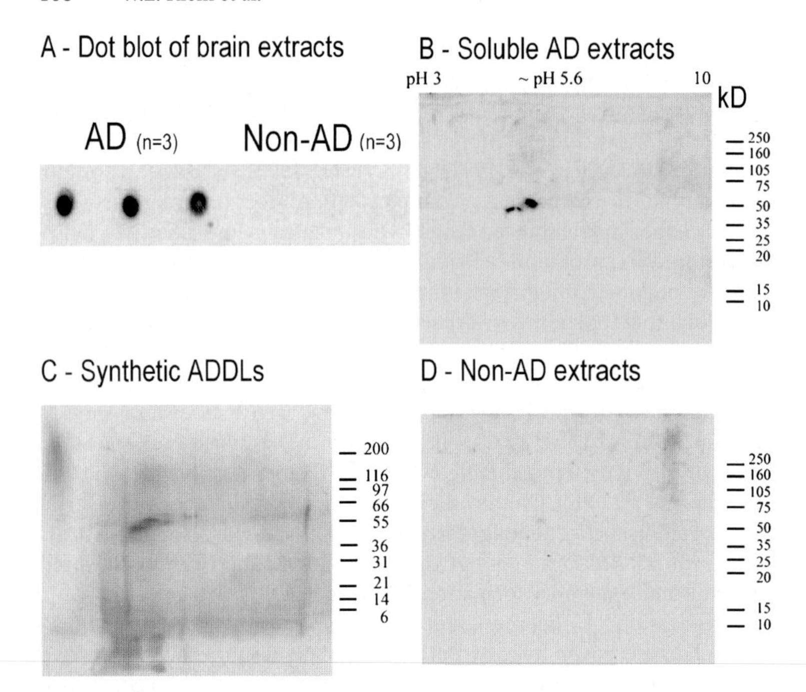

Fig. 3. ADDLs from AD brain are structurally identical to synthetic ADDLs. (**A**). Dot blots of AD-affected brain show large increases in Aβ oligomers compared to control brains, detected by our ADDL-selective antibody. (**B**) Immunoblot of human AD brain extract following 2D gel separation shows a predominant 12mer at pI 5.6, detected by ADDL-selective antibody. (**C**) Synthetic ADDLs also show a predominant 12mer (same size and isoelectric point as ADDLs from AD brain). (**D**) Non-AD brains had no detectable ADDLs (Gong et al. 2003)

antibodies that bind to assembled forms of Aβ but not monomers. More recently, monoclonal antibodies that target pathological assemblies but not monomers also have been generated using ADDLs as immunogen (Chromy et al. 2003; Lambert et al. 2007). These antibodies have different specificities for various oligomers and fibrils, but none bind robustly to the physiologically prevalent Aβ monomer. These conformation-dependent antibodies made it possible to develop highly sensitive dot-blot immunoassays for ADDLs in solution (Klein et al. 2000; Chang et al. 2001, 2003).

The application of dot-blot immunoassay to human brain extracts has confirmed the pathological relevance of ADDLs. ADDLs show a presence in human brain that is strikingly dependent on AD (Gong et al. 2003). Human brain homogenates prepared in detergent-free physiological buffer and clarified by ultracentrifugation show AD-affected brains accumulating ADDL levels up to 70-fold over controls. ADDL accumulation is high in extracts of frontal cortex but only marginally above background in cerebellum (Fig.3a; Lacor et al. 2004b), a pattern consistent with other aspects of AD pathology.

Structure of pathologically relevant oligomers: AD brain and animal models contain prominent 12mers. Three types of data indicate that the oligomers in AD brain are structurally and functionally equivalent to those studied in vitro. First, brain oligomers react with conformation-dependent antibodies raised against synthetic oligomers. Significantly, these antibodies not only distinguish oligomers from monomers but also distinguish toxic oligomers from non-toxic species (Lambert et al. 2001). Neither capacity is typical of antibodies generated against short peptides. Second, ADDLs from human brain mimic the ability of synthetic ADDLs to act as neuron-specific ligands and trigger neuronal pathology (discussed later). Third, synthetic and AD brain-derived preparations show equivalence by 2D-gel analysis. Each contains a prominent oligomer with an isoelectric point of ~5.6 and SDS-stable mass of ~54 kDa, i.e., 12mers (Fig.3b; Gong et al. 2003). Taken together, these results establish that the presence of ADDLs is a bona fide characteristic of AD brain pathology.

In harmony with human pathology, the occurrence of oligomers has been confirmed in multiple strains of tg-mouse AD models, with increases in immunoreactivity that reach several hundred-fold over controls (Chang et al. 2003). These robust increases in oligomers are regionally selective, indicating that broad diffusion does not occur. In the highly regarded triple transgene model of Oddo et al., which exhibits AD-like tau phosphorylation as well as plaques (Oddo et al. 2003), two different oligomer-selective antibodies showed the same pattern of developmental increase, with a transient peak in oligomers at six months and a sustained elevation after 15 months (Oddo et al. 2006). Another mouse model also showed highly upregulated soluble oligomers (Ohno et al. 2006). In these animals, knockout of β-secretase eliminated oligomers and blocked deficiencies in a fear-linked memory task.

In a widely recognized recent study, Ashe's group confirmed that tg-mice accumulate Aβ oligomers of the same size we previously identified in human AD brain (Lesne et al. 2006). Most importantly, this oligomer (which they call Aβ*56) correlates with memory loss. Aβ*56 is first detectable in Western blots at an age coinciding with the development of memory dysfunction. Since the mouse 12mer is derived from the human APP transgene, it is likely to be identical to the 12mer found in AD brain. Overall, the abundance of oligomers resulting from transgene expression in transgenic animal models of AD likely explains why brain deficiencies correlate poorly with amyloid plaques, and it substantiates their value as models for early AD pathology and memory loss.

It should be noted that evidence available as early as 1994 indicated the presence of pre-fibrillar Aβ assemblies in AD-affected brains (Kuo et al. 1996; Frackowiak et al. 1994). However, at that time only fibrils were regarded as the initiators of pathogenesis, so oligomers were taken to be surrogates for ongoing fibrillogenesis, simply transient intermediates en route to formation of the pathogenically relevant amyloid fibrils. As illustrated in Fig. 1, this point of view has changed significantly over the past decade.

How ADDLs and Neurons Interact: ADDLs are Gain-of-Function Pathogenic Ligands that Target Particular Synapses

ADDL distribution in situ. Knowing that ADDLs are abundant in AD brain extracts, we have used our conformation-dependent antibodies to investigate distribution in situ. As seen in Fig. 4, AD brain sections exhibit localized immunoreactivity that selectively surrounds cell bodies (Lacor et al. 2004a; Lambert et al. 2007). Distribution

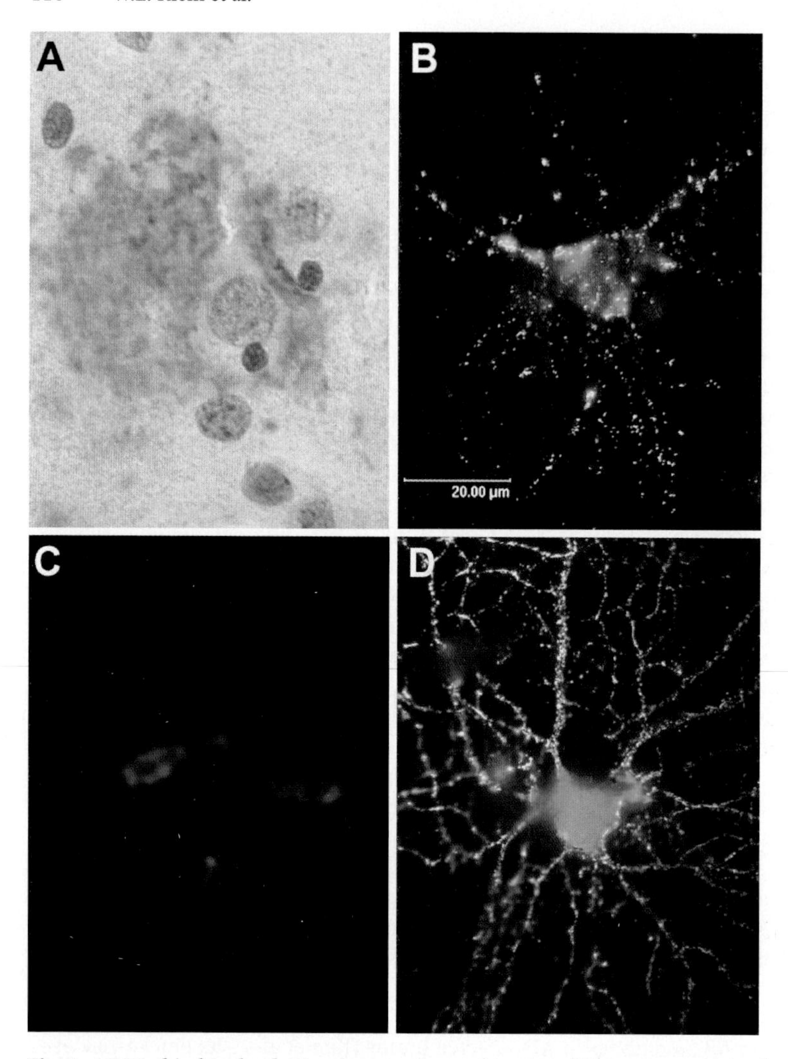

Fig. 4. ADDLs bind to dendritic arbors in vivo and in vitro. (**A**) In situ analysis of postmortem AD frontal cortex using our ADDL-selective antibody shows ADDLs are localized perineuronally within the dendritic arbor. (**B**) ADDLs from AD brain extracts bind to cultured hippocampal neurons at dendritic hot spots, mimicking their in situ distribution, whereas no binding was detectable with non-AD brain extract (**C**). (**D**) Synthetic ADDL binding is identical to AD brain-derived ADDLs (Gong et al. 2003; Lacor et al. 2004b)

is distinct from fibrillar amyloid deposits detected by thioflavin-S staining, a pattern first observed by Glabe, Kayed and colleagues using the oligomer-specific antibodies generated at their lab (Kayed et al. 2003). These results establish the in situ presence of oligomers distinct from fibrils, and they validate the conclusion obtained from immuno-dot-blot assays that ADDLs accumulate in AD brain.

Overall, oligomer staining in AD sections shows little intracellular localization. The pericellular distribution itself is reminiscent of the diffuse, synaptic-type deposit observed with prion-associated diseases (Hainfellner et al. 1997; Kovacs et al. 2002). The pericellular diffuse immunoreactivity is found in all AD cases and also is evident in some preclinical samples. The preclinical labeling occurred in the absence of other pathology, suggesting that pericellular accumulation of ADDLs is an early event in pathogenesis. The putative early association of oligomers with dendritic arbors suggests the appealing hypothesis that an attack on synapses by ADDLs may be the key reason why AD is memory specific.

ADDLs act as specific ligands. To test the hypothesis that pericellular labeling in situ represents the attachment of oligomers to specific dendritic targets, we carried out binding experiments using brain-derived ADDLs and highly differentiated cultures of rat hippocampal neurons (Gong et al. 2003; Lacor et al. 2004b). As predicted by the hypothesis, oligomers extracted from AD brain were found to attach to dendrites in culture (Fig. 4). Dendritic distribution shows a discrete punctate pattern, consistent with localization to synapses (Gong et al. 2003; Lacor et al. 2004b). Identical patterns of punctate binding were found for brain-derived and synthetic ADDLs, as expected from their structural equivalence. No binding was found in extracts from control non-AD subjects. Identical patterns of binding were evident in living as well as fixed cells, establishing that ADDL association with neurons occurred at the plasma membrane. Once bound to the membrane, it appears possible that ADDLs might subsequently be internalized.

Many neurons do not exhibit hot spots of ADDL binding. Cell-to-cell specificity in a double-label experiment was illustrated for a pair of αCaMKII-positive neurons, only one of which exhibited ADDL binding (Fig. 5; Lacor et al., 2004b). Over many experiments, the subpopulations of neurons that bind ADDLs typically comprised 30–50% of the total in a given culture. ADDLs thus do not bind to neurons through a mechanism involving non-selective membrane adsorption. Specific binding at the cellular level establishes that ADDLs essentially are gain-of-function pathogenic ligands.

A mechanism of action by specific ligand binding is predicated on the existence of ADDLs in extracellular compartments. Fulfillment of this requirement has been established by findings that ADDLs accumulate in cerebrospinal fluid (CSF). A sensitive, nanotechnology-based immunoassay showed that ADDLs in CSF are at least 10-fold higher in AD than in control samples (Georganopoulou et al. 2005; Haes et al. 2005). Because of the minimal overlap between control and patient levels, ADDLs in CSF have the potential to provide a significant biomarker for AD.

ADDL binding sites are coincident with a subset of synapses. At the subcellular level, whether ADDLs bind specifically to synapses is of significance to the hypothesis that memory loss in AD is an oligomer-induced synaptic failure (Lambert et al. 1998; Selkoe 2002). The rapidity with which ADDLs inhibit synaptic plasticity (Lambert et al. 1998; Walsh et al. 2002) suggests that the neurologically relevant binding might occur near synapses, and binding that specifically targets particular synapses would not only account for memory specificity in AD but would also confer considerable constraints on the mechanism. Localization of ADDLs to punctate binding sites within dendritic arbors clearly is consistent with the hypothesis that ADDLs are synapse-specific ligands.

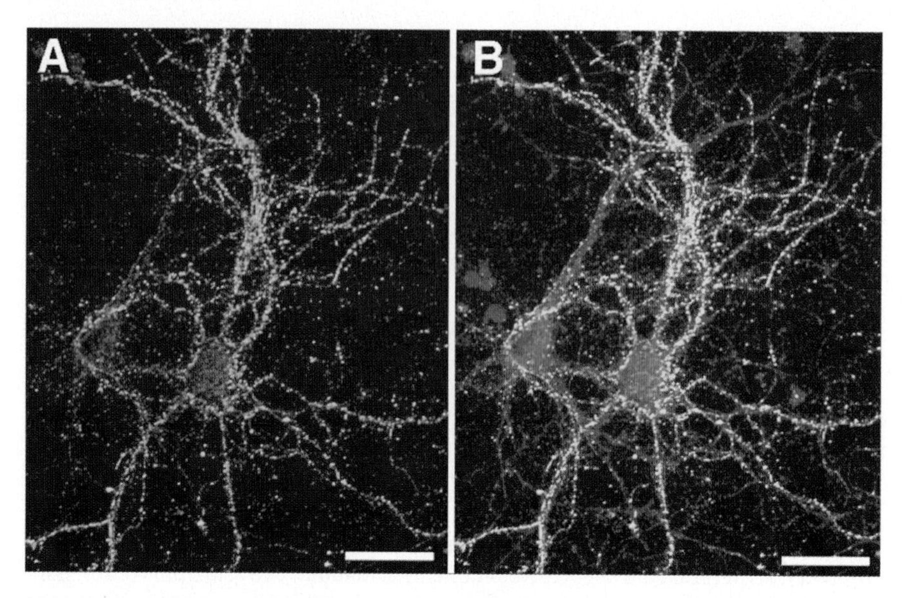

Fig. 5. ADDL binding is cell specific. Double-labeling immunocytochemistry was performed on ADDL-treated mature hippocampal neurons using rabbit polyclonal anti-ADDL (M94; **A**, *green*) and mouse monoclonal anti-αCaM kinase II (**B**, *red*). Overlay (**B**, *yellow*) shows that ADDLs bind selectively to some αCaMKII-positive neurons and, as depicted here, only one of the two neurons is targeted. *Scale bars*, 20 μm (Lacor et al. 2004b)

We have examined a possible colocalization between ADDLs and PSD-95 in highly differentiated rat hippocampal cultures (Lacor et al. 2004b), a preferred model for investigations of synaptic cell biology. PSD-95 is a critical scaffolding component of postsynaptic densities found in excitatory CNS signaling pathways (Sheng and Pak 1999), and clusters of PSD-95 represent definitive markers for postsynaptic terminals (Rao et al. 1998). As predicted, ADDL binding sites showed striking coincidence with PSD-95 puncta, as shown in an overlay in Fig. 6 (Lacor et al. 2004b). Overlay analyses indicated that the ADDL binding sites colocalized almost exclusively with puncta of PSD-95. Identical patterns were obtained with extracts of AD brain. The extent of colocalization between ADDLs and PSD-95 was quantified by image analysis and results showed that synthetic ADDLs colocalized with PSD-95 in $93 \pm 2\%$ of the sites. These sites were selective, furthermore, for a synaptic subpopulation: only half of the PSD-95 puncta colocalized with ADDL binding sites. The synapses targeted by ADDLs thus comprised approximately half of those that were present.

If synapses were targeted in situ, the impact on memory ultimately would depend on the number of synapses targeted, the extent to which individual synapses were compromised, and the relevance of affected synapses to the overall process of memory formation. In essence, memory loss depends on the tipping of balance between synapse failure due to ADDLs and the brain's synaptic reserve (Mesulam 1999), with no simple linear relationship between total ADDL abundance and memory dysfunction.

Sub-synaptic target of ADDLs: Dendritic spines. Dendritic spines are morphologically plastic protrusions, critical for synaptic information storage and memory, on which

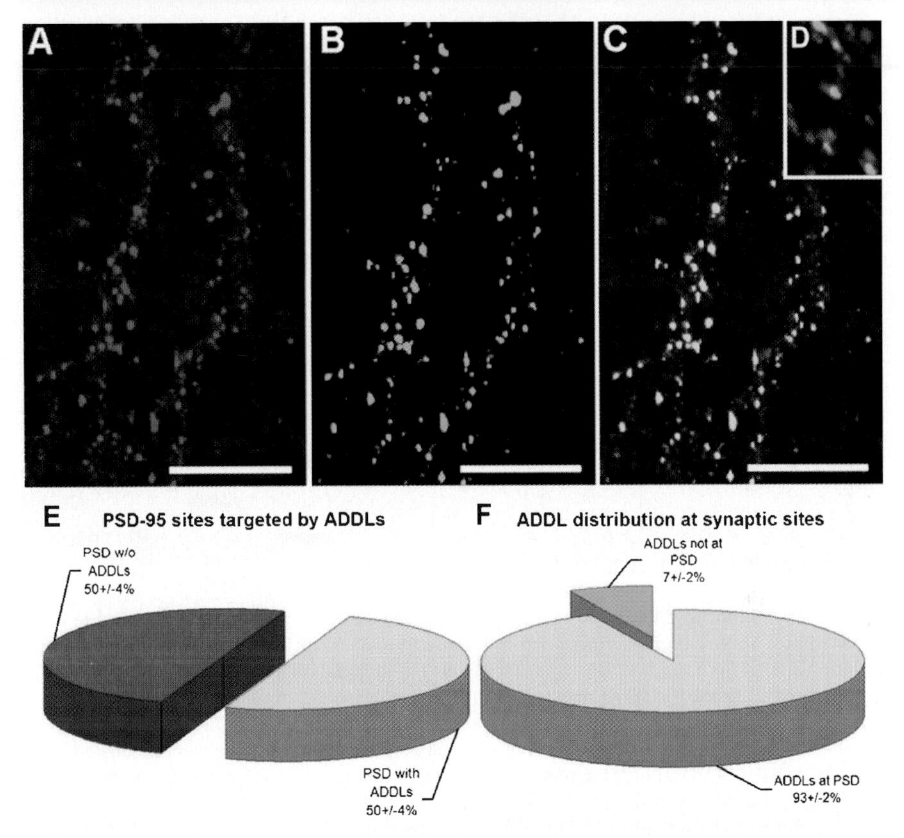

Fig. 6. ADDLs bind specifically to synapses. Mature hippocampal cultures incubated with 500 nM ADDLs and double labeled for PSD-95 (**A**, *red*) and ADDLs (**B**, *green*) show extensive coincidence for PSD-95 and ADDLs (**C**, *overlay, yellow*). Image analysis of ~15, 500 ADDL puncta and ~27, 500 PSD-95 puncta shows that half of the PSD sites are occupied by ADDLs (E) and that ADDL sites greatly (~93%) co-localize with PSD-95 (F). Analyses verify that ADDLs specifically localize to a subset of synaptic sites. *Scale bar*, 10 μm. (D) However, ADDLs (*green*) are juxtaposed to a presynaptic marker (synaptophysin in *red*) rather than overlapped as with a postsynaptic marker (PSD-95) (Lacor et al. 2004b)

the majority of excitatory inputs occur. Spines are specifically targeted by ADDLs, shown first by the colocalization of ADDLs with PSD-95 and with calcium/calmodulin-dependent kinase II-positive protrusions (Lacor et al. 2004b). Substantiation of this conclusion was provided by the recent immunofluorescence colocalization of ADDLs with drebrin (Lacor et al. 2007), a cytoskeletal marker known to be concentrated within spine heads (Aoki et al. 2005).

Biochemical fractionation confirms ADDL attachment to post-synaptic targets. In vitro, ADDLs bind to synaptosomes from forebrain but not cerebellum (Fig. 7; Lacor et al. 2007), a regional specificity consistent with previously observed ADDL toxicity (Kim et al. 2003). Detergent treatment yielded an ADDL binding complex distributing with postsynaptic densities but not with presynaptic active-zone proteins.

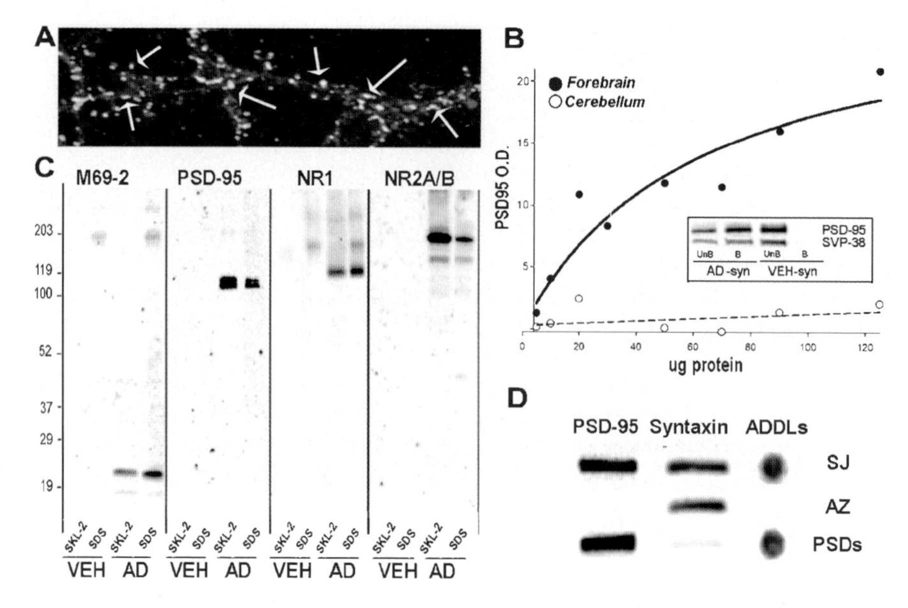

Fig. 7. ADDL binding sites are postsynaptic. (**A**) Image demonstrates that ADDLs bind to dendritic spines. Mature hippocampal neurons exposed to 500 nM ADDLs for 15 min were double immunolabeled for bound ADDLs (*cyan*) with our monoclonal oligomer-selective antibody and drebrin (*red*), a dendritic spine marker. Merged image demonstrates a high level of colocalization (*white, arrows*) between ADDL puncta and dendritic spines. (**B**) ADDL binding is region specific. Different amounts of forebrain or cerebellar crude synaptosomes were incubated with 500 nM ADDLs. ADDL-bound synaptosomes were immunoprecipitated using our polyclonal anti-oligomer antibody. Immunoprecipitated material was assayed for PSD-95. Increasing PSD-95 immunoreactivity was observed only from the forebrain synaptosomes. Inset, ADDL-treated (AD) or vehicle-treated (VEH) synaptosomes (syn) were immunoprecipitated using our ADDL antibody. Bound (**B**) and unbound (UnB) fractions were assessed for the presence of post- and pre-synaptic markers (PSD-95 and SVP-38). Only the "AD-syn" bound fraction showed immunoreactivity for both synaptic markers. (**C**) ADDL-treated synaptosomes (**AD**), labeled with anti-ADDLs, incubated with Triton X-100 and DOC, and immunoprecipitated with anti-mouse IgG show coisolation of ADDLs (labeled with M69-2) with postsynaptic proteins (PSD-95, NR1, and NR2A/B), all released in second Sarkosyl (SKL-2) and SDS detergent extraction, demonstrating that the ADDL binding complex segregates with postsynaptic markers. (**D**) Dissociation of presynaptic and postsynaptic compartments of ADDL-bound synaptosomes demonstrates that ADDLs cofractionate with synaptic junctions (SJ) and PSDs and not active zones (AZ) (Lacor et al. 2007)

These complexes can be isolated using magnetic immunobeads. Elution of ADDLs from isolated complexes also releases PSD-95 and the NR1 and NR2A/B subunits of NMDA receptors, markers ofpostsynaptic densities (Fig. 7; Lacor et al. 2007). The data verify that the ligand action of ADDLs targets components of synaptic spines.

Molecular domains of synaptic targeting – ADDLs attach to toxin receptors at or near NMDA receptors. The molecular basis for specific synaptic targeting by ADDLs is not known, although flow cytometry experiments have established that specific ADDL binding is mediated by trypsin-sensitive cell-surface proteins (Lambert et al. 1998).

In hippocampal cultures, ADDL binding sites are found on microtubule-associated protein-2-positive cells that express NMDA-R subunits NR1 and NR2B (Lacor et al. 2007). No clusters of bound ADDLs occur on astrocytes (GFAP-positive cells) or inhibitory neurons expressing GAD (Lacor et al. 2007). Binding site overlap with NMDA receptor (NR1) immunoreactivity is consistent with the typical association of PSD-95 and NMDA glutamate receptors in excitatory hippocampal signaling pathways (Sheng and Pak 1999). Although a wide range of candidate proteins accumulate at PSD-95-positive synapses, including receptors for neurotransmitters, trophic factors, adhesion molecules, and extracellular matrix proteins, we have found that an antibody directed at the extracellular domain of NR1 reduces ADDL binding by ~60% (De Felice et al. 2007a). These results indicate that ADDL binding at synaptic spines is in close proximity to NMDA-Rs. In accord with this hypothesis, the NR1 subunit of NMDA receptors co-immunoprecipitates with ADDLs from detergent-extracted, ADDL-treated rat synaptosomal membrane preparations (Lacor et al. 2007; De Felice et al. 2007a). The incomplete block of ADDL binding by NR1 antibodies indicates heterogeneity in binding domains. In addition, because NR1 subunits also occur on neurons showing little or no ADDL binding, this polypeptide in itself is insufficient for high ADDL binding. An appealing hypothesis is that the ADDL ligand may in fact bind to a domain comprising more than one specific protein target, essentially a molecular cassette of alternative components.

ADDL toxicity involves NMDA receptors – a disease-specific mechanism to explain therapeutic benefits of memantine (Namenda). The proximity of ADDL binding to NMDA-Rs suggests that these receptors could play a pivotal role in ADDL-instigated neuronal damage. This possibility has been confirmed in investigations of ADDL-induced ROS formation. ADDLs stimulate neuronal ROS formation well above physiological levels, consistent with the involvement of oxidative stress in AD, and this response is inhibited 100% by the same NR1 antibody that blocks ADDL binding. A C-terminal (intracellular) antibody used as a control is without effect (Fig. 8; De Felice et al. 2007a). It is noteworthy that the N-terminal antibody to the NR1 subunit of NMDA-Rs does not cause functional impairment of the receptor per se, as indicated by the fact that ROS formation induced by NMDA is unaffected by the antibody.

Memantine, an NMDA-R open-channel blocker, has recently been approved for AD treatment (prescribed as Namenda). Memantine modestly preserves memory in patients with moderate-to-severe forms of AD, and its beneficial actions have so far been related to the prevention of excessive NMDA-R activation (i.e., excitotoxicity; Lipton 2006). According to that view, the efficacy of memantine in neurological diseases associated with NMDA-R overactivation is related to prevention of the excessive influx of Ca2+ through the receptor's associated ion channel, leading to ROS formation (Lipton 2006). However, it is not sufficiently clear why a drug that blocks NMDA-Rs, which play essential roles in synaptic plasticity and LTP, might be beneficial in terms of preserving memory in AD patients (Morris 2001; Schmitt 2005; reviewed in Nakazawa et al. 2004).

We investigated the hypothesis that memantine might protect against neuronal damage initiated by soluble Aβ oligomers acting at NMDA-Rs. At doses comparable to its clinical pharmacological window, memantine potently inhibited ADDL-induced ROS formation in hippocampal neurons (De Felice et al. 2007a), suggesting that one

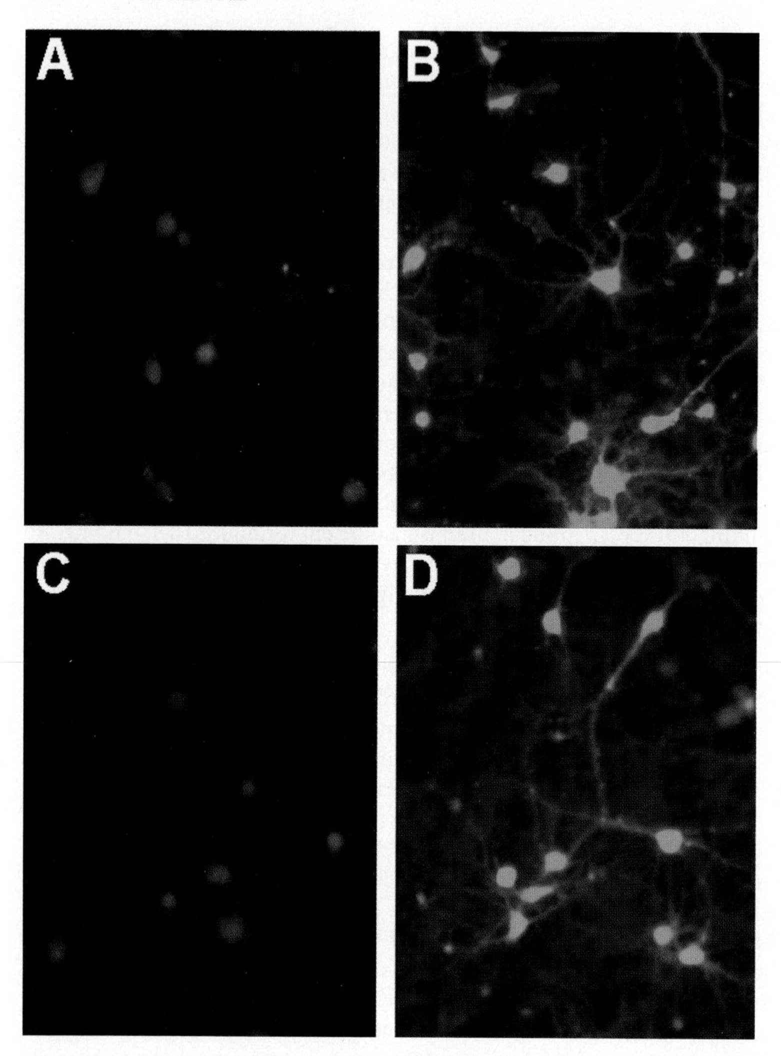

Fig. 8. ADDL-stimulated ROS formation is blocked by anti-NR1. Mature hippocampal cultures were treated with 500 nM ADDLs or vehicle for 4 hours and ROS levels were measured. (**A**) Vehicle-treated cultures showed minimal ROS. (**B**) ADDL treatment greatly stimulated ROS production. (**C**) ROS elevation was blocked by antibodies against the N-terminal epitope of NMDA-R1 subunit, but not by C-terminal epitope of NMDA-R1 (**D**) (De Felice et al. 2007a)

possible mechanism of action of memantine involves prevention of neuron damage due to excessive ADDL-induced glutamatergic stimulation. Consistent with this possibility, memantine also exhibits efficacy in blocking ADDL-induced drebrin loss and insulin receptor down-regulation (see below). The cognitive benefits conferred to AD patients by memantine may thus derive from protection against the deleterious synaptic impact of ADDLs.

Other NMDA-R antagonists, MK-801 and APV, also block ADDL-induced ROS formation (De Felice et al. 2007a), but these are not useful for AD therapeutics because of their tight binding. Interestingly, APV (but not MK-801) significantly reduces ADDL binding to hippocampal neurons. Inhibition of ADDL binding by APV (which is consistent with inhibition by an anti-NR1 antibody) suggests that the APV coordination domain within the agonist binding site may either represent a direct ADDL binding domain or may be involved, via the induction of protein conformational changes, in the regulation of ADDL interactions with the NMDA-R or with closely associated protein receptors.

It is interesting to note that memantine itself does not block ADDL binding to hippocampal neurons (De Felice et al. 2007a), which may account for the clinical failure of memantine with time, as ADDLs presumably would trigger further neuron damage whenever the memantine dissociated from the receptor pore. New drugs optimized as antagonists of ADDL activity potentially could provide improved AD therapeutics.

ADDL-Induced AD Pathology: Tau Hyperphosphorylation and Aberrant Synapse Composition, Shape, and Abundance

Link to tau-pathology in cell biology and animal models. In addition to oxidative damage, ADDLs stimulate tau phosphorylation at epitopes characteristically hyperphosphorylated in AD (De Felice et al, 2007b). Significantly, tau hyperphosphorylation in culture also is induced by a soluble aqueous extract containing Aβ oligomers from AD brain tissue, but not by an extract from non-AD brains. Conformation-dependent antibodies block attachment of brain-derived ADDLs to synaptic binding sites and prevent tau hyperphosphorylation. Tau phosphorylation is also blocked by the Src family tyrosine kinase inhibitor, 4-amino-5-(4-chlorophenyl)-7(t-butyl)pyrazol(3,4-d)pyramide (PP1), and by the phosphatidylinositol-3-kinase inhibitor, LY294002. These findings confirm the putative link between ADDLs and tau pathology indicated earlier in studies of the triple transgenic mouse developed by LaFerla, Oddo and colleagues (Oddo et al. 2006). When these animals were injected with the oligomer-specific antibody of Glabe, tau pathology was greatly reduced. These results from cell biology and animal models strongly support the hypothesis that ADDL activity is directly linked to tau pathology in AD.

Ectopic expression of Arc, a memory-linked IEG. Localization of ADDL binding sites to dendritic spines suggests a potential for rapid impact on spine molecules. In fact, within minutes of binding, ADDLs have been found to ectopically upregulate the synaptic memory-linked immediate-early gene (IEG) protein "Arc" (Fig. 9; Lacor et al. 2004a, b). Arc (*activity-regulated cytoskeletal-associated*) mRNA is targeted to synapses where, physiologically, the protein normally is induced transiently by synaptic activity (Link et al. 1995; Lyford et al. 1995; Steward and Worley 2001). Following a short-term exposure of hippocampal neurons to ADDLs, oligomer binding colocalizes with dendritic punctate Arc expression (Lacor et al. 2004b). Binding at this location appears to be an ectopic induction, because low levels of Arc protein that are expressed constitutively localize in cell bodies and not at synapses (Steward and Worley 2001). After longer exposure to ADDLs, the expression of Arc exhibits a robust upregulation (Lacor

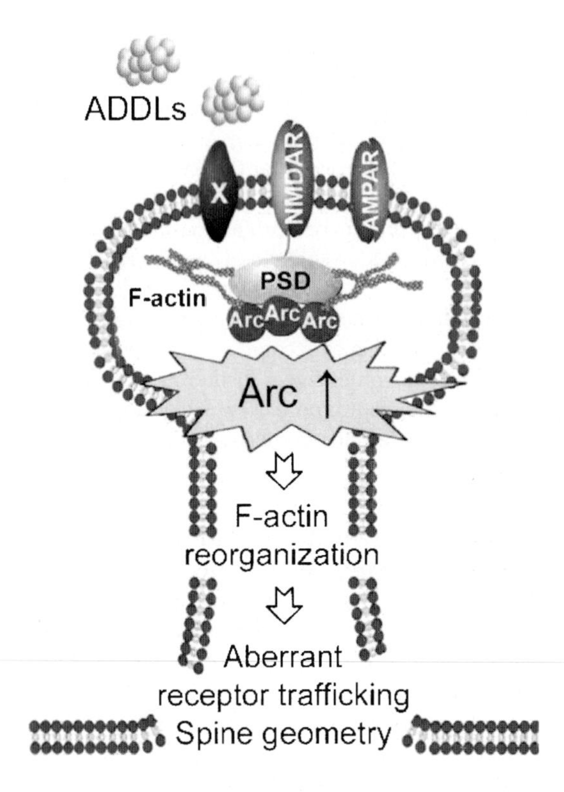

Fig. 9. Predicted consequences of ADDL attack on synapses: disruption of receptor trafficking and aberrant dendritic spine structure. Cytoskeletal anomalies involving ectopic arc over-expression led us to predict that ADDLs would cause aberrations in spine shape and receptor trafficking (Lacor et al. 2004b), subsequently verified by our group (Lacor et al. 2007) and others (Snyder et al. 2005, 2006; Hsieh et al. 2006; Shankar et al. 2007)

et al., 2004b), five-fold over vehicle-treated cultures. Interestingly, other investigators focusing on Arc itself had earlier predicted that ectopic and aberrant expression of this IEG would cause failure of long-term memory formation, a prediction validated in mice where Arc expression was repressed or over-expressed (Guzowski et al. 2000; Kelly and Deadwyler 2003).

Verified predictions – ADDLs disrupt receptor trafficking and spine shape. Arc protein is known to regulate the state of F-actin. The potential for an irregular cytoskeleton led us to predict that ADDLs would cause aberrations in spine shape and receptor trafficking, giving a cell biological basis for synapse failure and memory loss (Fig. 9; Lacor et al. 2004b). With respect to receptors, given that ADDLs block onset of LTP (Lambert et al., 1998) and reversal of LTD (Wang et al. 2002) and disrupt NMDA-receptor-mediated CREB phosphorylation (Tong et al. 2001), we proposed that a likely consequence would be reduction in receptors for glutamate (Gong et al. 2003). This prediction has been confirmed by our own group (Lacor et al., 2004c, 2005, 2007) as well as others (Snyder et al. 2005; Hsieh et al. 2006). After 3 h of ADDL treatment, both NR1 and NR2B surface expression decreases dramatically (by 78% and 70%, respectively; Lacor et al. 2007), potentially providing a cellular mechanism for the significant loss of NMDA-R in AD brain (Sze et al. 2001; Mishizen-Eberz et al. 2004). We also tested the effect of ADDLs on EphB2 receptors, which can be physically coupled to NMDA-R at synaptic sites (Dalva et al. 2000). Like NMDA-R, EphB2 signaling is germane to mechanisms of synaptic plasticity (Matynia et al. 2002; Carlisle and Kennedy 2005). In neurons exposed to ADDLs, EphB2 showed a major decrease, exhibiting a 60% decrease in EphB2 puncta after 6 h (Lacor et al. 2007).

The large ADDL-induced decrease in receptor expression occurs before any major change in spine density, consistent with synaptic plasticity being compromised before degeneration, and suggests that changes in surface NMDA targeting/cycling and anchoring in this brain region may play a role in synaptic plasticity failure associated with AD at early stages.

With respect to spine morphology, Kelly and Deadwyler proposed that elevated Arc expression in Arc tg-mice, which interferes with learning, causes spines to become more rigid (Kelly and Deadwyler 2003), consistent with the association of Arc with cytoskeletal and postsynaptic proteins (Lyford et al. 1995; Fujimoto et al. 2004). Changes in spine structure may rapidly alter synaptic signal processing and related information storage (Crick 1982; Rao and Craig 2000; Bonhoeffer and Yuste 2002), and spine abnormalities are common to various brain dysfunctions (Fiala et al. 2002), including mental retardation, in which spines are atypically bent and protruding (Kaufmann and Moser 2000).

In response to ADDLs, spines manifest major restructuring. We first noted that control spines are stubby and lie close along dendritic shafts, whereas ADDL-treated spines are longer and appear to extend out from the dendrites (Lacor et al., 2004b, c). This altered shape has been confirmed and extended using spines immunolabeled for drebrin (Lacor et al. 2007), an excellent marker that localizes to the cortical cytoskeleton of dendritic spines forming excitatory synapses (Allison et al. 2000; Aoki et al. 2005). In the presence of ADDLs, dendrites show abnormally elongated protrusions rather than the distinctive headed protrusions characteristic of mature dendritic spines (Fig. 10). Initial time points show short spines with heads close to the shaft, characteristic of mature (stubby and mushroom types) spines. Thin filopodia-like morphology develops with continued ADDL exposure time. Shelanski and colleagues, moreover, have found that acute slice preparations exposed to Aβ also exhibit abnormally elongated spines (Shrestha et al. 2006). Change in spine morphology caused by ADDLs is especially interesting because the elongated shape resembles that of immature spines or of the diseased spines found in mental retardation and prionoses (Fiala et al. 2002).

Degeneration and elimination of spines. The morphological reorganization initiated by ADDLs ultimately culminates in a loss of dendritic spines. Drebrin-labeled spines in neurons with bound ADDLs show a striking, time-dependent decrease in density. At 3 h of exposure, spine density is not significantly changed, but by 6 h, a decrease of 33% is observed, and by 24 h, the decrease in spines is 50% (Lacor et al. 2007). Neurons not targeted by ADDLs displayed normal spine density and shape, indicating that the mechanism of spine degeneration in AD is instigated at the molecular level by the binding of ADDLs to synaptic terminals. The drebrin loss is marked and itself has the

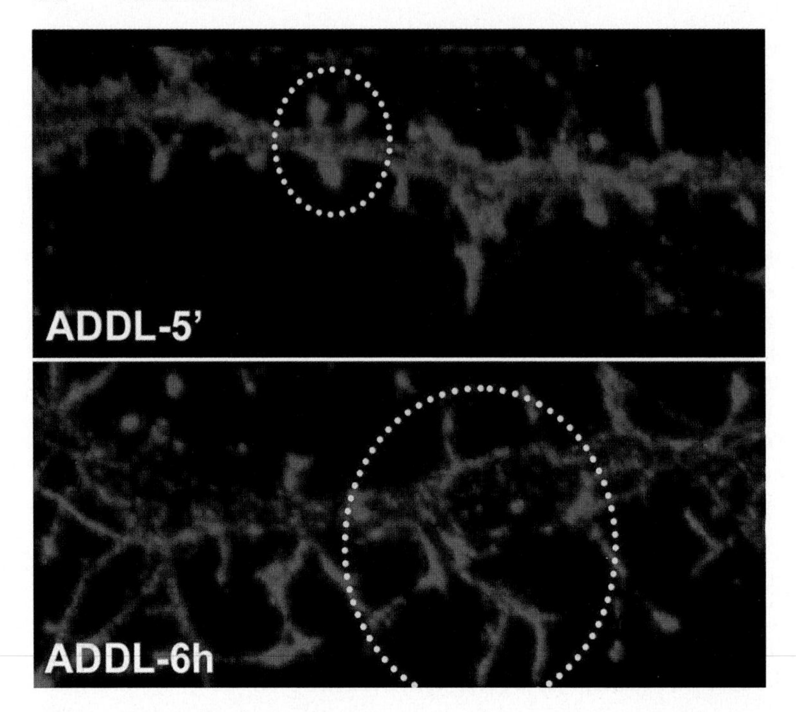

Fig. 10. ADDL-induced spine elongation. Mature hippocampal neuron cultures were incubated with 500 nM ADDLs or vehicle and immunolabeled for drebrin, an actin binding protein. By 6 hr, ADDLs induced a striking spine elongation (see *circles*) compared to short incubation time where spines were short and stubby (Adapted from Lacor et al. 2007)

potential to be part of the degenerative mechanism, as drebrin plays a physiological role in actin assembly and clustering of PSD-95 (Takahashi et al. 2004).

Pathological loss of spines and their associated molecules is well documented for AD brain (Scheibel and Tomiyasu 1978; Shim and Lubec 2002; Scheff and Price 2003) and tg mouse AD models (Lanz et al. 2003; Calon et al. 2004; Moolman et al. 2004; Spires et al. 2005; Jacobsen et al. 2006). The identity of molecules that cause spine deterioration in situ is not yet established, but, because both AD subjects (Gong et al. 2003; Lacor et al. 2004b) and tg-mice accumulate substantial levels of ADDLs (Chang et al. 2003; Oddo et al. 2006; Ohno et al. 2006), it has been proposed that these synaptic ligands are the responsible toxins (Lacor et al. 2004b; Klein 2006). This hypothesis is consistent with the observations from Mucke and colleagues that synapse loss in tg-mice is fibril-independent (Hsia et al. 1999; Mucke et al. 2000), although others have noted that spine loss in mouse models appears to occur near prominent plaques (Spires et al. 2005; Grutzendler et al. 2007). Tg mice, however, also exhibit decreased drebrin (Calon et al. 2004), consistent with loss caused by ADDLs in culture.

The significant decrease in the number of synaptic terminals in AD brain has been suggested to be the best correlate of dementia (DeKosky and Scheff 1990; Terry et al. 1991; Sze et al. 1997), and the known relationship between abnormal spine morphology and cognitive dysfunction implies that ADDL-induced spine pathologies could be a major factor in the loss of functional connectivity in AD.

Ligand-size dependence of drebrin loss – putative role for 12mers. We have tested the relationship between oligomer size, ligand binding, and drebrin loss using ADDLs fractionated by 10, 50 kDa and 100 kDa filters (Lacor et al. 2004b, 2007). Overall results demonstrate that soluble Aβ species that bind to dendritic processes and induce synapse loss comprise $A\beta_{1-42}$ oligomers between 50 and 100 kDa. Retentates of greater than 50 kDa show characteristic punctate binding on dendritic processes and lead to severe drebrin loss (Lacor et al. 2007). In contrast, $A\beta_{1-40}$ monomers, as well as species that might have formed during the 24 h incubation with cells, did not affect drebrin expression (Lacor et al. 2007). With respect to size, these synaptotoxic oligomeric ligands are consistent with the prominent 12mers found in AD brain (Gong et al. 2003) and with a 56 kDa Aβ* species reported for a tg AD mouse model (Lesne et al. 2006). Oligomer-induced loss of spines also has been observed by Shankar, Selkoe and colleagues in experiments using the 3 and 4mer preparations found in cell-conditioned medium (Shankar et al. 2007). The metabolically produced oligomers do not elicit changes in spine shape and may assume a conformation distinct from those found in synthetic solutions.

Because the pathogenic consequences of ADDLs derive from a ligand-like attachment to specific synaptic proteins (Gong et al. 2003; Lacor et al. 2004b), it will be important in the future to identify these toxin receptors and determine their role in inducing the cellular substrates of dementia. Novel therapeutic strategies aimed at blocking ADDL attachment to synapses or contravening its consequences should ultimately be successful in preserving cognitive function by preventing the critical pathological loss of synaptic connectivity.

ADDLs Provide a Basis for CNS Insulin Resistance in AD

Down-regulation of insulin receptors. Very recently, hippocampal neurons have been found to respond to low-dose ADDLs with a striking loss of dendritic insulin receptors (IRs; Zhao et al. 2007). In these cultured neurons, IRs show a punctate dendritic distribution consistent with the synaptic occurrence of IRs reported for brain tissue (Zhao et al. 1999) and in harmony with the ability of insulin to rapidly affect mechanisms of synaptic plasticity (Wan et al. 1997; Mielke and Wang 2005; Biessels et al. 1996). ADDLs cause a major loss of receptors from these puncta in the dendritic plasma membrane by 30 minutes (Fig. 11). Double-labeling shows that dendrites with bound ADDLs (green) are essentially devoid of IRs (red); in contrast, IRs are abundant on dendrites without bound ADDLs. Because in control cultures robust IR immunoreactivity occurs on all dendrites (40 of 40 neurons inspected), it is impossible that ADDLs are targeting only neurons lacking insulin receptors. By image analysis, the magnitude of insulin receptor decrease is at least 70%. Significantly, neurons affected by ADDLs showed elevated IR levels in their cell bodies, suggesting a rapid redistribution of IRs rather than net receptor loss. This conclusion is supported by Western blots that showed no ADDL-induced changes in total IR levels. The rapid redistribution of IRs is consistent with reports in which other proteins important for synaptic plasticity, including NMDA

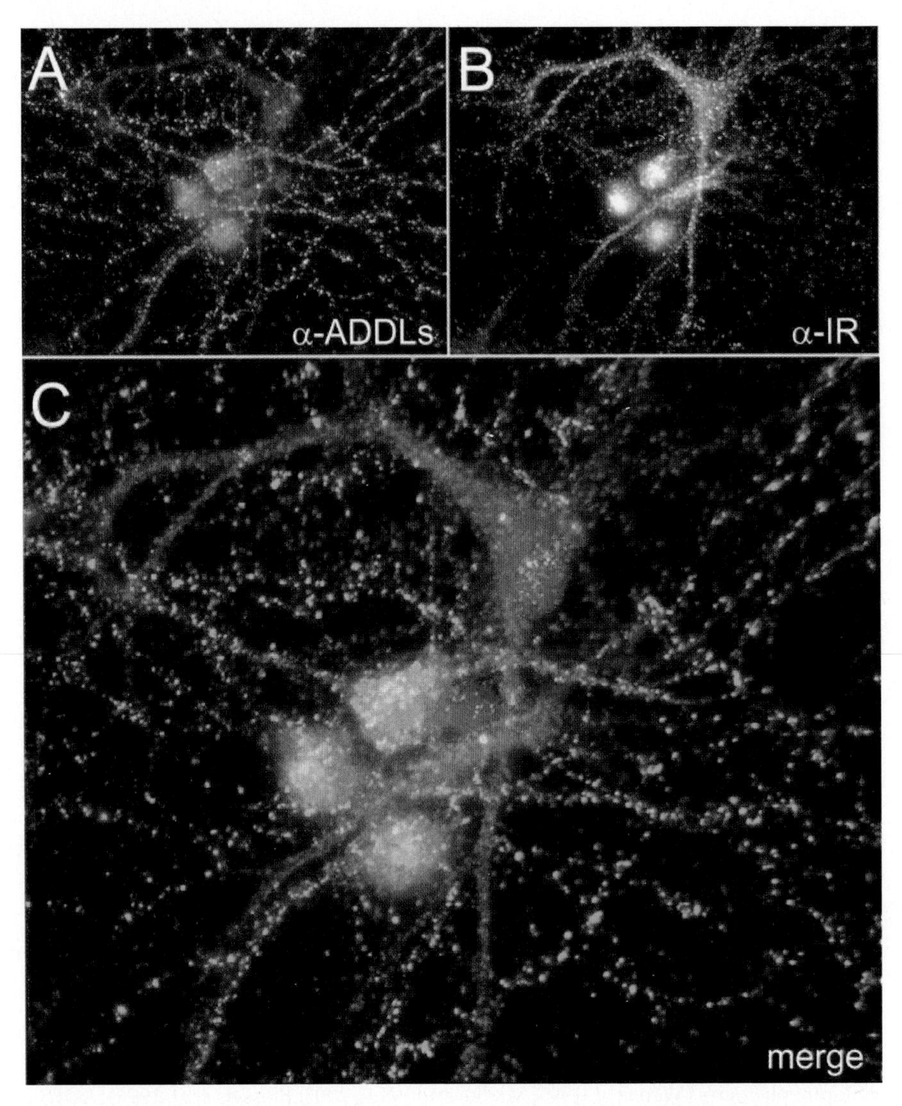

Fig. 11. ADDLs cause major redistribution of neuronal insulin receptors. Cultured hippocampal neurons were treated with 100 nM ADDLs for 30 min, followed by double staining for insulin receptor (IRα) and ADDL binding. Representative images showing ADDL binding detected by our anti-ADDL antibody (**A**) and the same field for the IRα (**B**). Merged images (**C**) show that ADDLs greatly reduce IRs from dendrites and promote IR accumulation in cell bodies. DAPI labeling (*blue*) indicates the nucleus (Zhao et al. 2007)

receptor subunits and EphB2 receptor tyrosine kinase, also showed surface loss caused by soluble forms of Aβ (Snyder et al. 2005; Lacor et al. 2007).

Decreased receptor responsiveness to insulin. Consistent with removal of IRs from dendritic membranes, neuronal responses to insulin are greatly lowered by ADDL

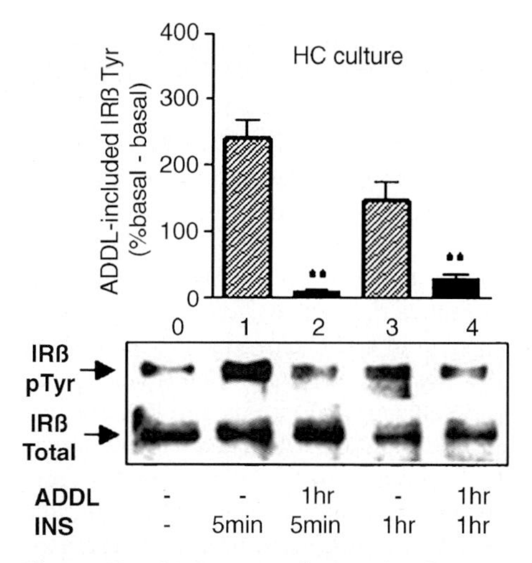

Fig. 12. IRs of CNS neurons become insulin resistant after short exposure to ADDLs. Hippocampal neurons were starved (medium sans B27) for 3–4 hr prior to 1-hr treatment with 50 nM ADDLs or vehicle in the presence and absence of insulin (100 nM). IR was immunoprecipitated with an IRβ antibody and the pTyr was detected using with 4G10 and Py20 (anti-pTyr antibodies). Lane 0: basal condition; Lanes 2–4: treated with insulin and/or ADDLs.
In the absence of ADDLs, stimulation of cells with insulin for 5 minutes caused a large increase in IR autophosphorylation. After 60 minutes, IR autophosphorylation was still robust but not as large. When ADDLs were present, the level of autophosphorylation due to insulin was greatly reduced (>80%), demonstrating that CNS neurons exposed to ADDLs develop major deficiencies in their response to insulin (Zhao et al. 2007)

treatment (Zhao et al. 2007). Our experiments have focused on the impact of ADDLs on insulin-induced receptor protein tyrosine kinase activity, measured by receptor autophosphorylation. In the absence of ADDLs, stimulation of cells with insulin for 5 minutes causes a large increase in IR autophosphorylation (Fig. 12). After 60 minutes, IR autophosphorylation is still robust but not as large. When ADDLs are present, however, the level of autophosphorylation due to insulin is reduced >80% in both hippocampal and cortical cultures (Fig. 12), demonstrating that CNS neurons exposed to ADDLs develop major deficiencies in their response to insulin (Zhao et al. 2007).

The possibility that ADDLs might directly affect IR function is consistent with previous results that kinase activity of semi-purified IRs is inhibited by $A\beta_{1-40}$ (Xie et al. 2002). However, the inhibitory dose of $A\beta_{1-40}$ in these experiments was 50 μM, at least 1000-fold greater than that used for ADDL incubation. Whether ADDLs directly interact with IRs is not yet clear. Significantly, since some cells with abundant insulin receptors show no ADDL binding (Fig. 11), the insulin receptor per se does not constitute a high-affinity ADDL binding site. However, co-immunoprecipitation of ADDLs with IRs suggests the presence of ADDLs and IRs in a multi-protein complex. Support for possible direct IR-ADDL interaction comes from transfected 3T3 cells, used because they provide a cell model with very high IR levels. These receptors, which respond to ADDLs with decreased autophosphorylation, co-immunoprecipitate with ADDLs and bind ADDLs in ligand blot assays (Zhao et al. 2007). The interactions appear to depend on IR tyrosine phosphorylation, as they are markedly increased by prior stimulation of cells with low doses of insulin. ADDL interactions with the IRs of 3T3 cells are relatively specific, as ADDL treatments did not result in binding to IGF-1 receptors or affect IGF-1 receptor activity (Zhao et al. 2007). Overall, the data indicate that ADDLs bind to a post-synaptic receptor complex that includes insulin receptors, leading to ADDL-induced CNS neuronal insulin resistance.

Mechanistically, NMDA-R signaling appears to play a role in down-regulation of IR function. Glutamate and depolarization reduce the responsiveness of IRs to insulin, likely involving Ca2+ influx and activation of Ca2+-dependent kinases (Zhao et al. 2007). These effects are in harmony with existing IR regulatory mechanisms and suggest a possible physiological feedback between neuronal activity and IR signaling. Inhibition is more severe when ADDL-treated neurons are stimulated with glutamate and depolarization. Significantly, inhibition is prevented by the NMDA blockers, memantine and APV, indicating NMDA receptors may mediate the observed IR inhibition. We also found that chelation of Ca2+ with BAPTA-AM prevented IR inhibition caused by ADDL treatments. The basis for Ca2+ dependent insulin resistance may involve pSer/pThr of the receptor by protein kinase C or cAMP-dependent kinase, a known intra-molecular negative regulatory factor for IR activity (Pessin and Saltiel 2000; Bossenmaier et al. 1997; Considine et al. 1995; Avignon et al., 1996). Protein kinase C in brain cells showed activation and membrane translocation triggered by Aβ (Nakai et al. 2001; Tanimukai et al. 2002) and ADDLs (Stevens 1998).

Insulin receptor dysfunction and loss of LTP. Independent of our investigations, Selkoe and colleagues carried out experiments that also led to the conclusion that Aβ oligomers cause dysfunctional IR activity. Their studies, using metabolically produced oligomers, are described elsewhere in this book. Most importantly, their new findings indicate that oligomer-induced IR dysfunction is a major factor in the mechanism that leads to failure of LTP.

ADDL-induced phosphorylation of Akt at Ser473. ADDLs also affect Akt, a critical signaling molecule of the IR pathway downstream of insulin receptor substrate (IRS) and PI3 kinase. During normal insulin signaling, Akt is first activated by phosphorylation at threonine308, which stimulates glucose transport and cell survival events (Morisco et al. 2005). With prolonged activation, Akt becomes phosphorylated at serine473, which is a key event in a negative feedback loop that inhibits IR signaling. Akt responds to ADDLs with increased phosphorylation of Ser473 (Zhao et al. 2007). Stimulation of Akt-pSer473 occurs at low levels of ADDLs and happens whether or not insulin is present, suggesting involvement of a pathway independent of IRs. Consistent with the action of ADDLs, stimulation of Akt-pSer473 in neurons is associated with NMDA receptor activation (Grimble 2002; Perkinton et al. 2002). Persistent elevation of Akt-pSer473 in brain appears to be undesirable, as abnormally enhanced Akt Ser473 phosphorylation is evident in AD brain (Griffin et al. 2005; Rickle et al. 2004), Huntington's disease striatal cells (Gines et al. 2003) and the brains of the Niemann-Pick type C (NPC) disease animal model (Rickle et al. 2004). Two targets negatively regulated by Akt-pSer473 are IRs and PI3K, which become threonine phosphorylated (Tian 2005). Significantly, high levels of Akt-pSer473 are associated with inflammation and peripheral insulin resistance diseases (Pessin and Saltiel 2000; Le Roith and Zick 2001; Lee and White 2004), suggesting the possibility that elevated Akt-pSer473 induced by Aβ oligomers could contribute to insulin resistance in AD-affected brain.

Significance of IR dysfunction. Our recent results show that ADDLs compromise insulin signaling in cultured brain cells, while parallel studies from Selkoe and colleagues have shown that metabolically derived Aβ oligomers also cause IR dysfunction, in a manner that underlies failure of synaptic plasticity. It is likely that ADDLs accumulating in AD brain would exert similar effects. The specific consequences of ADDL-induced brain insulin resistance with respect to cognitive function are difficult to predict, given that knowledge of the precise roles of CNS IRs is at present limited. Various reports have indicated memory-enhancing effects of insulin, learning-associated changes in IR pathways, and impairments of memory and LTP in diabetic animals (Wan et al. 1997; Mielke and Wang 2005; Zhao and Alkon 2001). Although specific deletion of the brain insulin receptor reportedly causes no obvious learning and memory impairment (Schubert et al. 2004), the absence of phenotype can be difficult to interpret. The receptor-deficient mice do show tau hyperphosphorylation, a major aspect of AD neuropathology (Schubert et al. 2004), possibly involving dysfunction of GSK3β and downstream molecules such as mTOR and p70 S6 kinase.

Impairment of insulin receptor signaling appears to be an important aspect of the overall synaptic pathology induced by ADDLs. It suggests that insulin resistance in AD brain is a response to ADDLs, which disrupt insulin signaling and may cause a brain-specific form of diabetes (Zhao et al. 2007). Damage to this pathway along with other plasticity-associated receptors seems likely to play an important role in the synaptic failure considered to be the mechanistic basis for the memory dysfunction of AD.

Recent experiments have shown that ADDL binding to neurons can be blocked by preincubation with high doses (1 μM) of insulin, consistent with the possible involvement of IRs as co-receptors. In addition to effectively blocking ADDL binding, insulin prevents ADDL-induced down-regulation of dendritic IRs. The conclusion is that IRs are necessary but not sufficient for high-affinity ADDL binding. A simple hypothesis is that ADDLs bind to a receptor complex that includes IRs and a heterologous "co-receptor" that is not present in all neurons. Therefore, it is possible that insulin may have important therapeutic implications in AD, preventing ADDL binding and blocking the pathology of insulin receptors.

Can ADDLs Explain why AD is Specific for Memory?

After 10 years of study, we now know that ADDLs represent a significant alternative to amyloid fibrils as the CNS neurotoxins responsible for the memory loss of AD. These soluble Aβ oligomers are long-lived neurotoxins, not simply intermediates in fibrillogenesis, and they accumulate in the CNS of individuals afflicted by AD. The oligomers found in AD brain, in mouse models, and in vitro show structural equivalence, although details regarding alternative-sized toxic oligomers and conformation states remain under active investigation.

The structure of ADDLs enables them to bind with specificity to particular synapses, mainly to synaptic spines at or near NMDA receptors. In essence, ADDLs are gain-of-function pathogenic ligands, and this capacity for specific binding provides a putative mechanism to explain why AD is a disease of memory. Binding disrupts synaptic plasticity, causes over-expression of the memory-linked immediate early gene Arc, and triggers pathological changes in synapse shape, composition, and survival. The attack on synapses provides a plausible mechanism unifying memory dysfunction with major features of AD neuropathology; recent findings show that ADDL binding also instigates synapse loss, oxidative damage, and AD-type tau hyperphosphorylation. Acting as novel neurotoxins that putatively account for memory loss and neuropathology, ADDLs present significant targets for disease-modifying therapeutics in AD, with proof-of-concept already evident from animal models.

References

Allison DW, Chervin AS, Gelfand VI, Craig AM (2000) Postsynaptic scaffolds of excitatory and inhibitory synapses in hippocampal neurons: maintenance of core components independent of actin filaments and microtubules. J Neurosci 20:4545–4554.

Aoki C, Sekino Y, Hanamura K, Fujisawa S, Mahadomrongkul V, Ren Y, Shirao T (2005) Drebrin A is a postsynaptic protein that localizes in vivo to the submembranous surface of dendritic sites forming excitatory synapses. J Comp Neurol 483:383–402.

Avignon A, Yamada K, Zhou X, Spencer B, Cardona O, Saba-Siddique S, Galloway L, Standaert ML, Farese RV (1996) Chronic activation of protein kinase C in soleus muscles and other tissues of insulin-resistant type II diabetic Goto-Kakizaki (GK), obese/aged, and obese/Zucker rats. A mechanism for inhibiting glycogen synthesis. Diabetes 45:1396–1404.

Bennett MR (2000) The concept of long term potentiation of transmission at synapses. Prog Neurobiol 60:109–137.

Biessels GJ, Kamal A, Ramakers GM, Urban IJ, Spruijt BM, Erkelens DW, Gispen WH (1996) Place learning and hippocampal synaptic plasticity in streptozotocin-induced diabetic rats. Diabetes 45:1259–1266.

Bitan G, Lomakin A, Teplow DB (2001) Amyloid beta-protein oligomerization: prenucleation interactions revealed by photo-induced cross-linking of unmodified proteins. J Biol Chem 276:35176–35184.

Bonhoeffer T, Yuste R (2002) Spine motility. Phenomenology, mechanisms, and function. Neuron 35:1019–1027.

Bossenmaier B, Mosthaf L, Mischak H, Ullrich A, Haring HU (1997) Protein kinase C isoforms beta 1 and beta 2 inhibit the tyrosine kinase activity of the insulin receptor. Diabetologia 40:863–866.

Boutaud O, Montine TJ, Chang L, Klein WL, Oates JA (2006) PGH-derived levuglandin adducts increase the neurotoxicity of amyloid beta1-42. J Neurochem 96:917–923.

Burdick D, Soreghan B, Kwon M, Kosmoski J, Knauer M, Henschen A, Yates J, Cotman C, Glabe C (1992) Assembly and aggregation properties of synthetic Alzheimer's A4/beta amyloid peptide analogs. J Biol Chem 267:546–554.

Calon F, Lim GP, Yang F, Morihara T, Teter B, Ubeda O, Rostaing P, Triller A, Salem N, Jr., Ashe KH, Frautschy SA, Cole GM (2004) Docosahexaenoic acid protects from dendritic pathology in an Alzheimer's disease mouse model. Neuron 43:633–645.

Carlisle HJ, Kennedy MB (2005) Spine architecture and synaptic plasticity. Trends Neurosci 28:182–187.

Chang L, Lambert MP, Viola KL, Gong Y, Venton DL, Krafft GA, Finch CE, Klein WL (2001) Non fibrillar Abeta toxins in AD: An immunoassay to characterize ADDL formation and identify ADDL-blocker compounds. Soc Neurosci.Abs 27:322.

Chang L, Bakhos L, Wang Z, Venton DL, Klein WL (2003) Femtomole immunodetection of synthetic and endogenous Amyloid-β oligomers and its application to Alzheimer's Disease drug candidate screening. J Mol Neurosci 20:305–313.

Chromy BA, Nowak RJ, Lambert MP, Viola KL, Chang L, Velasco PT, Jones BW, Fernandez SJ, Lacor PN, Horowitz P, Finch CE, Krafft GA, Klein WL (2003) Self-assembly of Abeta(1-42) into globular neurotoxins. Biochemistry 42:12749–12760.

Considine RV, Nyce MR, Allen LE, Morales LM, Triester S, Serrano J, Colberg J, Lanza-Jacoby S, Caro JF (1995) Protein kinase C is increased in the liver of humans and rats with non-insulin-dependent diabetes mellitus: an alteration not due to hyperglycemia. J Clin Invest 95:2938–2944.

Crick F (1982) Do dendritic spines twitch? Trends Neurosci 5:44–46.

Dalva MB, Takasu MA, Lin MZ, Shamah SM, Hu L, Gale NW, Greenberg ME (2000) EphB receptors interact with NMDA receptors and regulate excitatory synapse formation. Cell 103:945–956.

De Felice FG, Velasco PT, Lambert MP, Viola K, Fernandez SJ, Ferreira ST, Klein WL (2007a) Abeta oligomers induce neuronal oxidative stress through an N-methyl-D-aspartate receptor-dependent mechanism that is blocked by the Alzheimer drug memantine. J Biol Chem 282:11590–11601.

De Felice FG, Wu D, Lambert MP, Fernandez SJ, Velasco PT, Lacor PN, Bigio EH, Jerecic J, Acton PJ, Shughrue PJ, Chen-Dodson E, Kinney GG, Klein WL (2007b) Alzheimer's disease-type neuronal tau hyperphosphorylation induced by Abeta oligomers. Neurobiol Aging. Epub.

DeKosky ST, Scheff SW (1990) Synapse loss in frontal cortex biopsies in Alzheimer's disease: correlation with cognitive severity. Ann Neurol 27:457–464.

Dodart JC, Bales KR, Gannon KS, Greene SJ, DeMattos RB, Mathis C, DeLong CA, Wu S, Wu X, Holtzman DM, Paul SM (2002) Immunization reverses memory deficits without reducing brain Abeta burden in Alzheimer's disease model. Nature Neurosci 5:452–457.

Ferreira ST, Vieira MN, De Felice FG (2007) Soluble protein oligomers as emerging toxins in Alzheimer's and other amyloid diseases. IUBMB Life 59:332–345.

Fiala JC, Spacek J, Harris KM (2002) Dendritic spine pathology: cause or consequence of neurological disorders? Brain Res Brain Res Rev 39:29–54.

Frackowiak J, Zoltowska A, Wisniewski HM (1994) Non-fibrillar beta-amyloid protein is associated with smooth muscle cells of vessel walls in Alzheimer disease. J Neuropathol Exp Neurol 53:637–645.

Fujimoto T, Tanaka H, Kumamaru E, Okamura K, Miki N (2004) Arc interacts with microtubules/microtubule-associated protein 2 and attenuates microtubule-associated protein 2 immunoreactivity in the dendrites. J Neurosci Res 76:51–63.

Georganopoulou DG, Chang L, Nam JM, Thaxton CS, Mufson EJ, Klein WL, Mirkin CA (2005) Nanoparticle-based detection in cerebral spinal fluid of a soluble pathogenic biomarker for Alzheimer's disease. Proc Natl Acad Sci USA 102:2273–2276.

Gines S, Ivanova E, Seong IS, Saura CA, MacDonald ME (2003) Enhanced Akt signaling is an early pro-survival response that reflects N-methyl-D-aspartate receptor activation in Huntington's disease knock-in striatal cells. J Biol Chem 278:50514–50522.

Glabe CG (2006) Common mechanisms of amyloid oligomer pathogenesis in degenerative disease. Neurobiol Aging 27:570–575.

Glenner GG, Wong CW (1984) Alzheimer's disease: initial report of the purification and characterization of a novel cerebrovascular amyloid protein. Biochem Biophys Res Commun 120:885–890.

Gong Y, Chang L, Viola KL, Lacor PN, Lambert MP, Finch CE, Krafft GA, Klein WL (2003) Alzheimer's disease-affected brain: presence of oligomeric A beta ligands (ADDLs) suggests a molecular basis for reversible memory loss. Proc Natl Acad Sci USA 100:10417–10422.

Griffin RJ, Moloney A, Kelliher M, Johnston JA, Ravid R, Dockery P, O'Connor R, O'Neill C (2005) Activation of Akt/PKB, increased phosphorylation of Akt substrates and loss and altered distribution of Akt and PTEN are features of Alzheimer's disease pathology. J Neurochem 93:105–117.

Grimble RF (2002) Inflammatory status and insulin resistance. Curr Opin Clin Nutr Metab Care 5:551–559.

Grutzendler J, Helmin K, Tsai J, Gan WB (2007) Various dendritic abnormalities are associated with fibrillar amyloid deposits in Alzheimer's disease. Ann NY Acad Sci 1097:30–39.

Guzowski JF, Lyford GL, Stevenson GD, Houston FP, McGaugh JL, Worley PF, Barnes CA (2000) Inhibition of activity-dependent arc protein expression in the rat hippocampus impairs the maintenance of long-term potentiation and the consolidation of long-term memory. J Neurosci 20:3993–4001.

Haass C, Selkoe DJ (2007) Soluble protein oligomers in neurodegeneration: lessons from the Alzheimer's amyloid beta-peptide. Nature Rev Mol Cell Biol 8:101–112.

Haes AJ, Chang L, Klein WL, Van Duyne RP (2005) Detection of a biomarker for Alzheimer's disease from synthetic and clinical samples using a nanoscale optical biosensor. J Am Chem Soc 127:2264–2271.

Hainfellner JA, Liberski PP, Guiroy DC, Cervenakova L, Brown P, Gajdusek DC, Budka H (1997) Pathology and immunocytochemistry of a kuru brain. Brain Pathol 7:547–553.

Hardy J, Selkoe DJ (2002) The amyloid hypothesis of Alzheimer's disease: progress and problems on the road to therapeutics. Science 297:353–356.

Hardy JA, Higgins GA (1992) Alzheimer's disease: the amyloid cascade hypothesis. Science 256:184–185.

Hilbich C, Kisters-Woike B, Reed J, Masters CL, Beyreuther K (1991) Aggregation and secondary structure of synthetic amyloid beta A4 peptides of Alzheimer's disease. J Mol Biol 218:149–163.

Hsia AY, Masliah E, McConlogue L, Yu GQ, Tatsuno G, Hu K, Kholodenko D, Malenka RC, Nicoll RA, Mucke L (1999) Plaque-independent disruption of neural circuits in Alzheimer's disease mouse models. Proc Natl Acad Sci USA 96:3228–3233.

Hsieh H, Boehm J, Sato C, Iwatsubo T, Tomita T, Sisodia S, Malinow R (2006) AMPAR removal underlies Abeta-induced synaptic depression and dendritic spine loss. Neuron 52:831–843.

Jacobsen JS, Wu CC, Redwine JM, Comery TA, Arias R, Bowlby M, Martone R, Morrison JH, Pangalos MN, Reinhart PH, Bloom FE (2006) Early-onset behavioral and synaptic deficits in a mouse model of Alzheimer's disease. Proc Natl Acad Sci USA 103:5161–5166.

Katzman R, Terry R, DeTeresa R, Brown T, Davies P, Fuld P, Renbing X, Peck A (1988) Clinical, pathological, and neurochemical changes in dementia: a subgroup with preserved mental status and numerous neocortical plaques. Ann Neurol 23:138–144.

Kaufmann WE, Moser HW (2000) Dendritic anomalies in disorders associated with mental retardation. Cereb Cortex 10:981–991.

Kayed R, Head E, Thompson JL, McIntire TM, Milton SC, Cotman CW, Glabe CG (2003) Common structure of soluble amyloid oligomers implies common mechanism of pathogenesis. Science 300:486–489.

Kelly MP, Deadwyler SA (2003) Experience-dependent regulation of the immediate-early gene arc differs across brain regions. J Neurosci 23:6443–6451.

Kim HJ, Chae SC, Lee DK, Chromy B, Lee SC, Park YC, Klein WL, Krafft GA, Hong ST (2003) Selective neuronal degeneration induced by soluble oligomeric amyloid beta protein. FASEB J 17:118–120.

Klein WL (2000) Aβ toxicity in Alzheimer's Disease. In: Chesselet MF (ed) Molecular mechanisms of neurodegenerative diseases., Totowa, New Jersey: Humana Press, Inc., pp 1–49.

Klein WL (2002) ADDLs and protofibrils – the missing links? Neurobiol Aging 23:231–235.

Klein WL (2006) Synaptic targeting by Abeta oligomers (ADDLs) as a basis for memory loss in early AD. Alzheimer's and dementia 2:43–55.

Klein WL, Barlow A, Chromy B, Edwards C, Freed R, Lambert MP, Morgan TE, Rozovsky I, Trommer B, Viola KL, Wals P, Zhang C, Finch CE, Krafft GA (1997) "ADDLs" – Soluble Aβ oligomers that cause biphasic loss of hippocampal neuron function and survival. Soc Neurosci Abstr 23:1662.

Klein WL, Chromy B, Lambert MP, Tushan KL, Viola KL, Krafft GA, Finch CE (2000) Oligomer/conformation-dependent Abeta antibodies. Soc Neurosci Abst 26:1285.

Klein WL, Krafft GA, Finch CE (2001) Targeting small Abeta oligomers: the solution to an Alzheimer's disease conundrum? Trends Neurosci 24:219–224.

Kotilinek LA, Bacskai B, Westerman M, Kawarabayashi T, Younkin L, Hyman BT, Younkin S, Ashe KH (2002) Reversible memory loss in a mouse transgenic model of Alzheimer's disease. J Neurosci 22:6331–6335.

Kovacs GG, Zerbi P, Voigtlander T, Strohschneider M, Trabattoni G, Hainfellner JA, Budka H (2002) The prion protein in human neurodegenerative disorders. Neurosci Lett 329:269–272.

Kuo YM, Emmerling MR, Vigo-Pelfrey C, Kasunic TC, Kirkpatrick JB, Murdoch GH, Ball MJ, Roher AE (1996) Water-soluble Abeta (N-40, N-42) oligomers in normal and Alzheimer disease brains. J Biol Chem 271:4077–4081.

Lacor PN, Buniel MC, Cain PC, Chang L, Lambert MP, Klein WL (2004a) Synaptic targeting by Alzheimer's related Abeta oligomers. Neurobiol Aging 25[Suppl 2], S446:7–21.

Lacor PN, Buniel MC, Chang L, Fernandez SJ, Gong Y, Viola KL, Lambert MP, Velasco PT, Bigio EH, Finch CE, Krafft GA, Klein WL (2004b) Synaptic targeting by Alzheimer's-related amyloid beta oligomers. J Neurosci 24:10191–10200.

Lacor PN, Buniel MC, Klein WL (2004c) ADDLs (Aβ oligomers) alter structure and function of synaptic spines. 2004 Abstract Viewer/Itinerary Planner Washington, DC: Society for Neuroscience[Online], Program No. 218.3.

Lacor PN, Sanz-Clemente A, Viola KL, Klein WL (2005) Changes in NMDA receptor subunit 1 and 2B expression in ADDL-treated hippocampal neurons. 2005 Abstract Viewer/Itinerary Planner Washington, DC: Society for Neuroscience[Online], Program No. 786.17.

Lacor PN, Buniel MC, Furlow PW, Clemente AS, Velasco PT, Wood M, Viola KL, Klein WL (2007) Abeta oligomer-induced aberrations in synapse composition, shape, and density provide a molecular basis for loss of connectivity in Alzheimer's disease. J Neurosci 27:796–807.

Lambert MP, Stevens G, Sabo S, Barber K, Wang G, Wade W, Krafft G, Snyder S, Holzman TF, Klein WL (1994) Beta/A4-evoked degeneration of differentiated SH-SY5Y human neuroblastoma cells. J Neurosci Res 39:377–385.

Lambert MP, Barlow AK, Chromy BA, Edwards C, Freed R, Liosatos M, Morgan TE, Rozovsky I, Trommer B, Viola KL, Wals P, Zhang C, Finch CE, Krafft GA, Klein WL (1998) Diffusible, nonfibrillar ligands derived from Abeta1-42 are potent central nervous system neurotoxins. Proc Natl Acad Sci USA 95:6448–6453.

Lambert MP, Viola KL, Chromy BA, Chang L, Morgan TE, Yu J, Venton DL, Krafft GA, Finch CE, Klein WL (2001) Vaccination with soluble Abeta oligomers generates toxicity-neutralizing antibodies. J Neurochem 79:595–605.

Lambert MP, Velasco PT, Chang L, Viola KL, Fernandez S, Lacor PN, Khuon D, Gong Y, Bigio EH, Shaw P, De Felice FG, Krafft GA, Klein WL (2007) Monoclonal antibodies that target pathological assemblies of Abeta. J Neurochem 100:23–35.

Lanz TA, Carter DB, Merchant KM (2003) Dendritic spine loss in the hippocampus of young PDAPP and Tg2576 mice and its prevention by the ApoE2 genotype. Neurobiol Dis 13:246–253.

Le Roith D, Zick Y (2001) Recent advances in our understanding of insulin action and insulin resistance. Diabetes Care 24:588–597.

Lee YH, White MF (2004) Insulin receptor substrate proteins and diabetes. Arch Pharm Res 27:361–370.

Lesne S, Koh MT, Kotilinek L, Kayed R, Glabe CG, Yang A, Gallagher M, Ashe KH (2006) A specific amyloid-beta protein assembly in the brain impairs memory. Nature 440:352–357.

Lewis DA, Higgins GA, Young WG, Goldgaber D, Gajdusek DC, Wilson MC, Morrison JH (1988) Distribution of precursor amyloid-beta-protein messenger RNA in human cerebral cortex: relationship to neurofibrillary tangles and neuritic plaques. Proc Natl Acad Sci USA 85:1691–1695.

Link W, Konietzko U, Kauselmann G, Krug M, Schwanke B, Frey U, Kuhl D (1995) Somatodendritic expression of an immediate early gene is regulated by synaptic activity. Proc Natl Acad Sci USA 92:5734–5738.

Lipton SA (2006) Paradigm shift in neuroprotection by NMDA receptor blockade: memantine and beyond. Nature Rev Drug Discov 5:160–170.

Lorenzo A, Yankner BA (1994) Beta-amyloid neurotoxicity requires fibril formation and is inhibited by congo red. Proc Natl Acad Sci USA 91:12243–12247.

Lyford GL, Yamagata K, Kaufmann WE, Barnes CA, Sanders LK, Copeland NG, Gilbert DJ, Jenkins NA, Lanahan AA, Worley PF (1995) Arc, a growth factor and activity-regulated gene, encodes a novel cytoskeleton-associated protein that is enriched in neuronal dendrites. Neuron 14:433–445.

Masters CL, Beyreuther K (1988) Neuropathology of unconventional virus infections: molecular pathology of spongiform change and amyloid plaque deposition. Ciba Found Symp 135:24–36.

Matynia A, Kushner SA, Silva AJ (2002) Genetic approaches to molecular and cellular cognition: a focus on LTP and learning and memory. Annu Rev Genet 36:687–720.

Maynard CJ, Bush AI, Masters CL, Cappai R, Li QX (2005) Metals and amyloid-beta in Alzheimer's disease. Int J Exp Pathol 86:147–159.

Mesulam MM (1999) Neuroplasticity failure in Alzheimer's disease: bridging the gap between plaques and tangles. Neuron 24:521–529.

Mielke JG, Wang YT (2005) Insulin exerts neuroprotection by counteracting the decrease in cell-surface GABA receptors following oxygen-glucose deprivation in cultured cortical neurons. J Neurochem 92:103–113.

Mishizen-Eberz AJ, Rissman RA, Carter TL, Ikonomovic MD, Wolfe BB, Armstrong DM (2004) Biochemical and molecular studies of NMDA receptor subunits NR1/2A/2B in hippocampal subregions throughout progression of Alzheimer's disease pathology. Neurobiol Dis 15:80–92.

Moolman DL, Vitolo OV, Vonsattel JP, Shelanski ML (2004) Dendrite and dendritic spine alterations in Alzheimer models. J Neurocytol 33:377–387.

Morisco C, Condorelli G, Trimarco V, Bellis A, Marrone C, Condorelli G, Sadoshima J, Trimarco B (2005) Akt mediates the cross-talk between beta-adrenergic and insulin receptors in neonatal cardiomyocytes. Circ Res 96:180–188.

Morris RG (2001) Episodic-like memory in animals: psychological criteria, neural mechanisms and the value of episodic-like tasks to investigate animal models of neurodegenerative disease. Philos Trans R Soc Lond B Biol Sci 356:1453–1465.

Mucke L, Masliah E, Yu GQ, Mallory M, Rockenstein EM, Tatsuno G, Hu K, Kholodenko D, Johnson-Wood K, McConlogue L (2000) High-level neuronal expression of abeta 1-42 in wild-type human amyloid protein precursor transgenic mice: synaptotoxicity without plaque formation. J Neurosci 20:4050–4058.

Nakai M, Tanimukai S, Yagi K, Saito N, Taniguchi T, Terashima A, Kawamata T, Yamamoto H, Fukunaga K, Miyamoto E, Tanaka C (2001) Amyloid beta protein activates PKC-delta and induces translocation of myristoylated alanine-rich C kinase substrate (MARCKS) in microglia. Neurochem Int 38:593–600.

Nakazawa K, McHugh TJ, Wilson MA, Tonegawa S (2004) NMDA receptors, place cells and hippocampal spatial memory. Nature Rev Neurosci 5:361–372.

Oda T, Wals P, Osterburg HH, Johnson SA, Pasinetti GM, Morgan TE, Rozovsky I, Stine WB, Snyder SW, Holzman TF, Krafft GA, Finch CE (1995) Clusterin (apoJ) alters the aggregation of amyloid beta-peptide (A beta 1-42) and forms slowly sedimenting A beta complexes that cause oxidative stress. Exp Neurol 136:22–31.

Oddo S, Caccamo A, Shepherd JD, Murphy MP, Golde TE, Kayed R, Metherate R, Mattson MP, Akbari Y, LaFerla FM (2003) Triple-transgenic model of Alzheimer's disease with plaques and tangles: intracellular Abeta and synaptic dysfunction. Neuron 39:409–421.

Oddo S, Caccamo A, Tran L, Lambert MP, Glabe CG, Klein WL, LaFerla FM (2006) Temporal profile of amyloid-beta (Abeta) oligomerization in an in vivo model of Alzheimer disease. A link between Abeta and tau pathology. J Biol Chem 281:1599–1604.

Ohno M, Chang L, Tseng W, Oakley H, Citron M, Klein WL, Vassar R, Disterhoft JF (2006) Temporal memory deficits in Alzheimer's mouse models: rescue by genetic deletion of BACE1. Eur J Neurosci 23:251–260.
Perkinton MS, Ip JK, Wood GL, Crossthwaite AJ, Williams RJ (2002) Phosphatidylinositol 3-kinase is a central mediator of NMDA receptor signalling to MAP kinase (Erk1/2), Akt/PKB and CREB in striatal neurones. J Neurochem 80:239–254.
Pessin JE, Saltiel AR (2000) Signaling pathways in insulin action: molecular targets of insulin resistance. J Clin Invest 106:165–169.
Pike CJ, Walencewicz AJ, Glabe CG, Cotman CW (1991) In vitro aging of beta-amyloid protein causes peptide aggregation and neurotoxicity. Brain Res 563:311–314.
Pike CJ, Burdick D, Walencewicz AJ, Glabe CG, Cotman CW (1993) Neurodegeneration induced by beta-amyloid peptides in vitro: the role of peptide assembly state. J Neurosci 13:1676–1687.
Rao A, Craig AM (2000) Signaling between the actin cytoskeleton and the postsynaptic density of dendritic spines. Hippocampus 10:527–541.
Rao A, Kim E, Sheng M, Craig AM (1998) Heterogeneity in the molecular composition of excitatory postsynaptic sites during development of hippocampal neurons in culture. J Neurosci 18:1217–1229.
Rickle A, Bogdanovic N, Volkman I, Winblad B, Ravid R, Cowburn RF (2004) Akt activity in Alzheimer's disease and other neurodegenerative disorders. Neuroreport 15:955–959.
Rodgers AB (2005) Progress report on Alzheimer's disease 2004–2005. U.S. Department of Health and Human Services; National Institutes on Aging; National Institutes of Health.
Scheff SW, Price DA (2003) Synaptic pathology in Alzheimer's disease: a review of ultrastructural studies. Neurobiol Aging 24:1029–1046.
Scheibel AB, Tomiyasu U (1978) Dendritic sprouting in Alzheimer's presenile dementia. Exp Neurol 60:1–8.
Scheibel ME, Lindsay RD, Tomiyasu U, Scheibel AB (1975) Progressive dendritic changes in aging human cortex. Exp Neurol 47:392–403.
Schmitt HP (2005) On the paradox of ion channel blockade and its benefits in the treatment of Alzheimer disease. Med Hypotheses 65:259–265.
Schubert M, Gautam D, Surjo D, Ueki K, Baudler S, Schubert D, Kondo T, Alber J, Galldiks N, Kustermann E, Arndt S, Jacobs AH, Krone W, Kahn CR, Bruning JC (2004) Role for neuronal insulin resistance in neurodegenerative diseases. Proc Natl Acad Sci U S A 101:3100–3105.
Selkoe DJ (2002) Alzheimer's disease is a synaptic failure. Science 298:789–791.
Shankar GM, Bloodgood BL, Townsend M, Walsh DM, Selkoe DJ, Sabatini BL (2007) Natural oligomers of the Alzheimer amyloid-beta protein induce reversible synapse loss by modulating an NMDA-type glutamate receptor-dependent signaling pathway. J Neurosci 27:2866–2875.
Sheng M, Pak DT (1999) Glutamate receptor anchoring proteins and the molecular organization of excitatory synapses. Ann NY Acad Sci 868:483–493.
Shim KS, Lubec G (2002) Drebrin, a dendritic spine protein, is manifold decreased in brains of patients with Alzheimer's disease and Down syndrome. Neurosci Lett 324:209–212.
Shrestha BR, Vitolo OV, Joshi P, Lordkipanidze T, Shelanski M, Dunaevsky A (2006) Amyloid beta peptide adversely affects spine number and motility in hippocampal neurons. Mol Cell Neurosci 33:274–282.
Snyder EM, Nong Y, Almeida CG, Paul S, Moran T, Choi EY, Nairn AC, Salter MW, Lombroso PJ, Gouras GK, Greengard P (2005) Regulation of NMDA receptor trafficking by amyloid-beta. Nature Neurosci 8:1051–1058.
Spires TL, Meyer-Luehmann M, Stern EA, McLean PJ, Skoch J, Nguyen PT, Bacskai BJ, Hyman BT (2005) Dendritic spine abnormalities in amyloid precursor protein transgenic mice demonstrated by gene transfer and intravital multiphoton microscopy. J Neurosci 25:7278–7287.
Stevens, G. R. (1998) Signaling proteins in Alzheimer's disease: The possible roles of focal adhesion kinase, paxillin, and protein kinase C. 101–112. PhD Thesis – Northwestern University.

Steward O, Worley PF (2001) Selective targeting of newly synthesized Arc mRNA to active synapses requires NMDA receptor activation. Neuron 30:227–240.

Stine WB, Jr., Dahlgren KN, Krafft GA, LaDu MJ (2003) In vitro characterization of conditions for amyloid-beta peptide oligomerization and fibrillogenesis. J Biol Chem 278:11612–11622.

Sze C, Bi H, Kleinschmidt-DeMasters BK, Filley CM, Martin LJ (2001) N-Methyl-D-aspartate receptor subunit proteins and their phosphorylation status are altered selectively in Alzheimer's disease. J Neurol Sci 182:151–159.

Sze CI, Troncoso JC, Kawas C, Mouton P, Price DL, Martin LJ (1997) Loss of the presynaptic vesicle protein synaptophysin in hippocampus correlates with cognitive decline in Alzheimer disease. J Neuropathol Exp Neurol 56:933–944.

Takahashi RH, Almeida CG, Kearney PF, Yu F, Lin MT, Milner TA, Gouras GK (2004) Oligomerization of Alzheimer's beta-amyloid within processes and synapses of cultured neurons and brain. J Neurosci 24:3592–3599.

Tanimukai S, Hasegawa H, Nakai M, Yagi K, Hirai M, Saito N, Taniguchi T, Terashima A, Yasuda M, Kawamata T, Tanaka C (2002) Nanomolar amyloid beta protein activates a specific PKC isoform mediating phosphorylation of MARCKS in Neuro2A cells. Neuroreport 13:549–553.

Terry RD (1994) Neuropathological changes in Alzheimer disease. Prog Brain Res 101:383–390.

Terry RD, Masliah E, Salmon DP, Butters N, DeTeresa R, Hill R, Hansen LA, Katzman R (1991) Physical basis of cognitive alterations in Alzheimer's disease: synapse loss is the major correlate of cognitive impairment. Ann Neurol 30:572–580.

Tian R (2005) Another role for the celebrity: Akt and insulin resistance. Circ Res 96:139–140.

Tong L, Thornton PL, Balazs R, Cotman CW (2001) Beta-amyloid-(1-42) impairs activity-dependent cAMP-response element-binding protein signaling in neurons at concentrations in which cell survival is not compromised. J Biol Chem 276:17301–17306.

Vollers SS, Teplow DB, Bitan G (2005) Determination of Peptide oligomerization state using rapid photochemical crosslinking. Methods Mol Biol 299:11–18.

Walsh DM, Klyubin I, Fadeeva JV, Cullen WK, Anwyl R, Wolfe MS, Rowan MJ, Selkoe DJ (2002) Naturally secreted oligomers of amyloid beta protein potently inhibit hippocampal long-term potentiation in vivo. Nature 416:535–539.

Wan Q, Xiong ZG, Man HY, Ackerley CA, Braunton J, Lu WY, Becker LE, MacDonald JF, Wang YT (1997) Recruitment of functional GABA(A) receptors to postsynaptic domains by insulin. Nature 388:686–690.

Wang HW, Pasternak JF, Kuo H, Ristic H, Lambert MP, Chromy B, Viola KL, Klein WL, Stine WB, Krafft GA, Trommer BL (2002) Soluble oligomers of beta amyloid (1-42) inhibit long-term potentiation but not long-term depression in rat dentate gyrus. Brain Res 924:133–140.

Xie L, Helmerhorst E, Taddei K, Plewright B, Van Bronswijk W, Martins R (2002) Alzheimer's beta-amyloid peptides compete for insulin binding to the insulin receptor. J Neurosci 22:RC221.

Zhao W, Chen H, Xu H, Moore E, Meiri N, Quon MJ, Alkon DL (1999) Brain insulin receptors and spatial memory. Correlated changes in gene expression, tyrosine phosphorylation, and signaling molecules in the hippocampus of water maze trained rats. J Biol Chem 274:34893–34902.

Zhao WQ, Alkon DL (2001) Role of insulin and insulin receptor in learning and memory. Mol Cell Endocrinol 177:125–134.

Zhao WQ, De Felice FG, Fernandez S, Chen H, Lambert MP, Quon M, Krafft GA, Klein WL (2007) Amyloid beta oligomers induce impairment of neuronal insulin receptors. FASEB J, in press.

Synaptic Transmission Dynamically Modulates Interstitial Fluid Amyloid-β Levels

John R. Cirrito[1], *Floy R. Stewart*[2], *Steven Mennerick*[3], and *David M. Holtzman*[4]

Summary. Aggregation of amyloid-β (Aβ) within the extracellular space of the brain into soluble and insoluble forms is central to the pathogenesis of Alzheimer's disease. Aβ is produced in neurons by cleavage of the amyloid precursor protein (APP) subsequent to release into the brain interstitial fluid (ISF). While a substantial amount is known about APP processing, the mechanisms that regulate Aβ release and that modulate soluble ISF Aβ levels are less clear. Several studies have suggested that postsynaptic receptor activation can alter APP processing. In vitro studies have demonstrated that synaptic activity can modulate Aβ levels in the media of cultured neurons as well. We used in vivo microdialysis to assess ISF Aβ levels in awake, behaving mice while altering hippocampal synaptic activity both pharmacologically and electrically. Electrical stimulation increased neuronal activity and rapidly increased ISF Aβ levels. In contrast, when hippocampal synaptic activity was inhibited with agents such as TTX and tetanus toxin, ISF Aβ levels were significantly reduced within hours of treatment. Using acute brain slices, we demonstrate that synaptic vesicle cycling alone, in the absence of presynaptic or postsynaptic depolarization, drives the release of Aβ from neurons. Because Aβ is not localized within synaptic vesicles, it is likely that an event closely associated with vesicle fusion and exocytosis is the key link between synaptic transmission and Aβ release. We propose that synaptic vesicle membrane endocytosis results in more APP endocytosis, thereby increasing Aβ generation and subsequent Aβ release into the extracellular space. These data have important implications for understanding the relationship between synaptic activity and Aβ levels. This relationship will likely also inform our understanding of the pathogenesis and treatment of Alzheimer's disease.

Introduction

Accumulation of amyloid-β (Aβ) within the brain extracellular space is a key step in the pathogenesis and progression of Alzheimer's disease (AD). Insoluble amyloid plaques, composed primarily of Aβ, are one pathological hallmark of AD. In addition, growing evidence suggests that soluble Aβ oligomers within the brain interstitial fluid (ISF) may participate in the disease. In vitro experiments suggest that conversion of Aβ from a monomeric peptide to either an oligomeric or fibrillar form is concentration-dependent. Thus, higher concentrations of soluble, extracellular Aβ may lead directly

[1] Department of Psychiatry, Neurology, Molecular Biology and Pharmacology, Hope Center for Neurological Disorders, Washington University School of Medicine, St. Louis, MO 63110, USA

[2] Department of Neurology, Hope Center for Neurological Disorders, Washington University School of Medicine, St. Louis, MO 63110, USA

[3] Department of Psychiatry, Anatomy, and Neurolobiology, Hope Center for Neurological Disorders, Washington University School of Medicine, St. Louis, MO 63110, USA, email: cirritoj@neuro.wustl.edu

[4] Department of Neurology, Molecular Biology and Pharmacology, Hope Center for Neurological Disorders, Washington University School of Medicine, St. Louis, MO 63110, USA

Selkoe et al.
Synaptic Plasticity and the Mechanism of Alzheimer's Disease
© Springer-Verlag Berlin Heidelberg 2008

to potentially toxic and disruptive Aβ species. Aβ within the brain extracellular fluid is particularly important because it can contribute to plaque formation and growth and has the potential to be toxic or inhibit synaptic transmission. Consequently, understanding factors that regulate Aβ levels within the brain ISF has implications for disease pathogenesis and may suggest methods to modulate disease progression.

High Levels of Neuronal Activity Correlate with Aβ Deposition in Humans

While a substantial amount is known about how Aβ is generated, for instance the molecules and pathways involved in the cleavage of amyloid precursor protein (APP), there is little known about the cellular pathways involved in Aβ production and release. Several studies in humans suggest that neuronal activity may be linked to Aβ deposition as plaques. For instance, 10% of individuals with temporal lobe epilepsy (TLE) develop diffuse Aβ plaques throughout the temporal lobe at ages when AD pathology would otherwise be rare (Mackenzie and Miller 1994; Gouras et al. 1997). Although the epileptic tissue where the plaques occur is not necessarily normal, these individuals share dramatically elevated neuronal activity compared to activity in the same brain regions of normal individuals. Individuals as young as 30 years of age already exhibited diffuse Aβ plaques at autopsy. Interestingly, even though Aβ plaques are present, TLE does not confer a greater risk of developing AD, possibly because the type of plaques that form do not contain a high β-sheet structure.

Other studies linking neuronal activity with Aβ deposition come from recent brain imaging work. Studies by Buckner and colleagues demonstrate that brain areas that have a high level of basal metabolic activity are the same brain areas that are most vulnerable to Aβ deposition in human AD patients (Buckner et al. 2005). Fluorodeoxyglucose-positron emission tomography (FDG-PET) demonstrated that when individuals are cognitively idle, meaning that they are not focusing on a particular task, stereotypical patterns of brain regions are active, including prefrontal, lateral temporal, precuneus, and lateral parietal cortices. These areas comprise what is referred to as the "default state" brain regions (Raichle et al. 2001). When individuals begin performing a particular mental task (e.g., reading a word), the default state brain regions generally decrease in metabolic activity while different areas increase in activity. Over the course of a lifetime, it is likely that the default state brain regions are some of the most metabolically active areas in the brain (Gusnard et al. 2001). FDG-PET measures of metabolic activity are believed to be a close corollary of neuronal activity. Interestingly, a cortical map of the default state regions and a map of brain regions that develop Aβ pathology in AD are remarkably similar (Buckner et al. 2005), suggesting that, in humans, areas that have the most overall neuronal activity are particularly vulnerable to Aβ deposition.

Postsynaptic Activation Modulates APP Processing

In animal models, the relationship between synaptic transmission and Aβ levels has been demonstrated at a molecular level. Postsynaptic receptor activation leads to alteration of Aβ levels by modulating the activity of proteases involved in APP cleavage.

APP is a single transmembrane domain protein that is cleaved by BACE1 followed by the γ-secretase complex to produce Aβ. Once produced, Aβ is secreted from neurons into the brain extracellular space, where it is a normal component of ISF. Alternatively, α-secretase cleaves APP within the Aβ sequence, a cleavage that precludes Aβ production. Activation of muscarinic M1 acetylcholine receptors in animal models and in humans increases α-secretase cleavage of APP, thus decreasing Aβ levels in the brain and CSF (Nitsch et al. 2000; Beach et al. 2001; Caccamo et al. 2006). In contrast, NMDA receptor activation can inhibit α-secretase, thus increasing extracellular Aβ levels in dissociated neuronal cultures (Lesne et al. 2005). The effect of both M1 receptor and NMDA receptor-mediated modulation of APP processing and Aβ levels occurs over hours to days.

Neuronal activity also modulates extracellular Aβ levels in organotypic brain slices that overexpress mutated forms of human APP transgenically or following virally driven expression. Picrotoxin, a non-competitive $GABA_A$ receptor antagonist, increased synaptic activity and elevated extracellular Aβ levels by 70%, whereas decreasing transmission with tetrodotoxin (TTX) depressed Aβ levels by 50% (Kamenetz et al. 2003). These in vitro studies assessed Aβ levels in media after 24–72 hours of treatment. The picrotoxin-dependent increase in extracellular Aβ levels is accompanied by increased β-CTF levels in the brain slice, which may suggest a change in β-secretase cleavage of APP. Whether this effect is driven by postsynaptic or presynaptic mechanisms, however, remains unclear. While it is clear that postsynaptic effects can modulate APP processing, it remains possible that presynaptic effects can affect Aβ production and release as well.

Increased Synaptic Activity Elevates ISF Aβ Levels in Vivo

To determine the relationship between synaptic activity and Aβ in vivo, we utilized a microdialysis technique to measure ISF Aβ levels longitudinally in awake, behaving mice. Microdialysis probes with a 38 kilo-Dalton molecular weight cut-off membrane are stereotaxically implanted into the left hippocampus of Tg2576 APP transgenic mice. This size membrane permits recovery of small peptides, such as Aβ (molecular weight 4.4 kDa), from the brain extracellular space. Microdialysis enables serial sampling of ISF Aβ every 30 minutes for up to 36 hours (Cirrito et al. 2003, 2005). By attaching field potential recording electrodes to the microdialysis guide cannula, we are able to assess extracellular field potentials (depth EEG) at the same time and location that we assess Aβ levels. To study normal Aβ metabolism, we utilized three- to five-month-old Tg2576 mice at an age that is prior to Aβ plaque accumulation. Experiments on mice of this age permit us to evaluate the effect of synaptic activity on Aβ without complications from Aβ deposits.

To determine if increased synaptic activity alters ISF Aβ levels, we implanted Tg2576 mice with microdialysis probes and recording electrodes within the hippocampus, as well as an ipsilateral stimulating electrode into the perforant pathway (Fig. 1). High frequency electrical stimulation of the perforant pathway generates seizures within the hippocampal formation, evident in extracellular field potential recordings (Sloviter et al. 1996; Cirrito et al. 2005). Hippocampal seizures increased ISF Aβ levels by 125% within the first 30 minutes of stimulation compared to pre-stimulation Aβ levels (Fig. 2A). In fact, as long as seizures continued, ISF Aβ levels remained elevated.

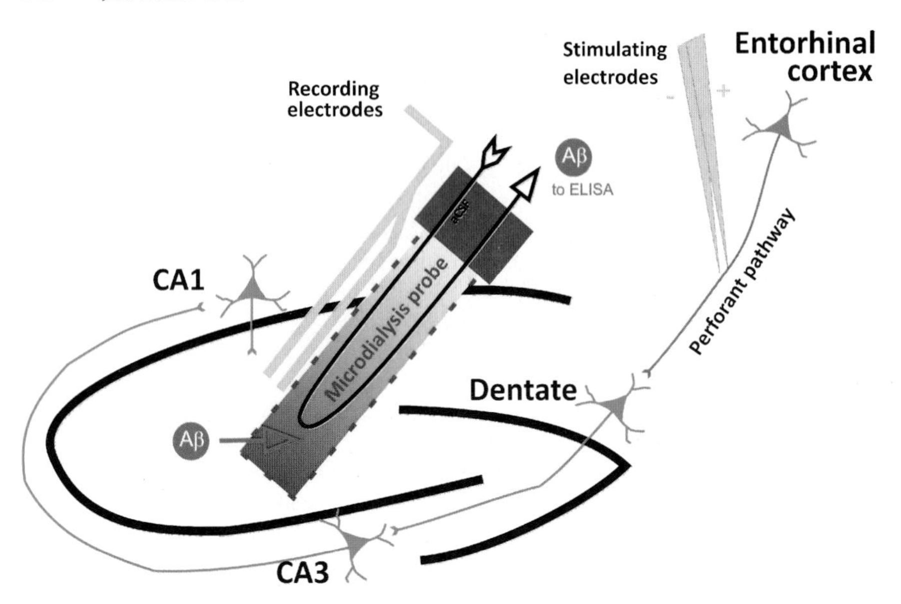

Fig. 1. Diagram of hardware implanted into mouse brain.
To measure ISF Aβ levels, a microdialysis probe (2 mm membrane length) was implanted into the hippocampus. Dual recording electrodes were attached to the guide cannula, enabling EEG recording at the dialysis location. Stimulating electrodes were also implanted into the perforant pathway, which, when stimulated at a high frequency, created seizures within the dentate gyrus and hippocampus (as confirmed with EEG recordings)

To determine if a more subtle increase in synaptic activity can also modulate ISF Aβ levels, we administered 25 μM picrotoxin directly into the hippocampus via reverse microdialysis. While high doses of picrotoxin can cause seizures (>200 μM), this low dose of the $GABA_A$ receptor antagonist caused occasional synchronous spikes in EEG activity but did not cause seizures (data not shown). Picrotoxin treatment increased ISF Aβ levels by 120% within the first 30 minutes of treatment. Levels reached a maximum increase of 145% by four hours compared to basal levels (Fig. 2A). Interestingly, a low dose of picrotoxin elevated ISF Aβ more than electrically stimulated seizures did. Stimulation-induced seizures caused brief bursts of very high levels of EEG activity (fast frequency and high amplitude) followed by interictal suppression of activity whereas picrotoxin caused a sustained, low level of elevated activity (spikes of spontaneous activity a 2–4 Hz). It is possible that the distinct kinetics and overall change in synaptic activity in these two paradigms account for the varying degrees of ISF Aβ alteration. Together, perforant pathway stimulation and picrotoxin treatment demonstrate that increased synaptic activity can cause a rapid and sustained increase in ISF Aβ levels.

Decreased Synaptic Activity Depresses ISF Aβ Levels in Vivo

As a complementary test of the relationship between synaptic activity and extracellular Aβ levels, we sought to determine if depressed activity would have the opposite

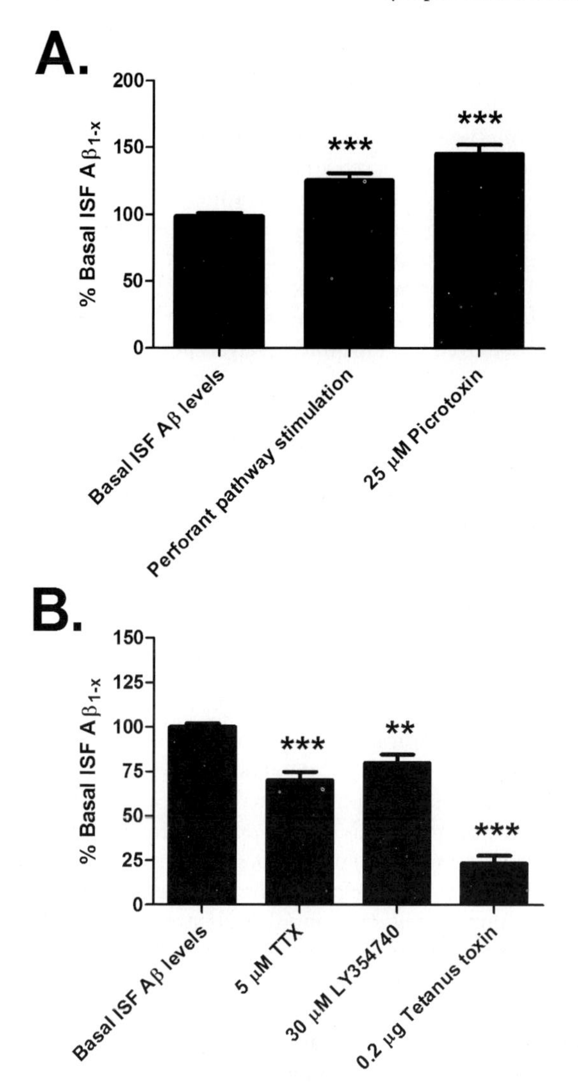

Fig. 2. Synaptic activity rapidly modulates ISF Aβ levels in vivo. (**A**) Electrical stimulation of the perforant pathway induced hippocampal seizures and increased ISF Aβ levels 125.5 ± 5.4% within 30 minutes compared to baseline ($p < 0.0001$, $n = 5$). Similarly, 25 μM picrotoxin via reverse microdialysis increased ISF Aβ levels by 145.4 ± 6.8% after four hours of treatment ($p < 0.0001$, $n = 3$). (**B**) Decreased synaptic transmission lowered ISF Aβ levels. Direct administration of TTX (5 μM) to the hippocampus gradually lowered ISF Aβ levels to 70.4 ± 4.5% after eight hours ($p < 0.0001$, $n = 5$). LY354740 (30 μM) inhibition of endogenous, glutamatergic synaptic transmission within the hippocampus decreased ISF Aβ levels by 80.0 ± 4.6% over six hours ($p < 0.01$, $n = 6$). Tetanus toxin (0.2 μg) blockade of synaptic vesicle exocytosis dramatically depressed ISF Aβ levels by 23.3 ± 4.6% by 18 hours after treatment ($p < 0.0001$, $n = 4$). Data expressed as mean ± SEM

effect of seizures and picrotoxin treatment. Tetrodotoxin (TTX) is a sodium channel blocker produced in puffer fish. When ingested in food or administered peripherally, TTX is lethal. If TTX is administered directly to the brain, it blocks action potential propagation and locally inhibits neuronal activity. After establishing basal ISF Aβ levels by microdialysis, mice were continuously administered TTX via reverse microdialysis directly into the hippocampus for eight hours. During this treatment, EEG activity within the hippocampus gradually declined to zero, with a concurrent decrease in ISF Aβ levels (Cirrito et al. 2005). By eight hours of treatment, ISF Aβ levels reached 70% of basal levels (Fig. 2B). The effect of TTX was reversible. Four hours of TTX treatment reduced EEG activity and Aβ levels; however both of these measures gradually returned to baseline levels when the drug was removed from the microdialysis perfusion buffer (Cirrito et al. 2005).

We also tested whether a manipulation of synaptic vesicle release modulates Aβ release. MGluR2/3 is expressed on the presynaptic terminals within the hippocampus. When activated, these receptors lower glutamate release by direct effects within the presynaptic terminal, thereby depressing synaptic transmission with little effect on overall cellular excitability. To determine if modulating presynaptic glutamate release alters ISF Aβ levels, we administered an agonist to metabotropic glutamate receptors 2 and 3 (mGluR2/3), LY354740, directly into the hippocampus. This drug decreased ISF Aβ levels by 20% over six hours of treatment, suggesting that endogenous synaptic activity contributes significantly to normal ISF Aβ levels (Fig. 2B).

TTX and LY354740 reduce, but do not completely prevent, neuronal activity. For instance, these drugs inhibit evoked synaptic transmission but do not block spontaneous activity. To block synaptic transmission more completely, we administered a bolus injection of tetanus toxin directly into the hippocampus surrounding a microdialysis probe. Tetanus toxin is taken into the presynaptic terminal and cleaves VAMP2, which is necessary for synaptic vesicle exocytosis. This process does not interfere with upstream action potential formation and propagation but does prevent exocytosis, thereby blocking all types of local synaptic activity. Tetanus toxin reduced the levels of ISF Aβ by almost 75% (Fig. 2B). This dramatic reduction in Aβ levels strongly suggests that synaptic vesicle exocytosis may be a critical event involved in Aβ release from cells.

Synaptic Vesicle Exocytosis is Critical for Aβ Release in Acute Brain Slices

Presynaptic neurotransmission involves fusion of the synaptic vesicle with the plasma membrane, release of neurotransmitter into the synaptic cleft, and subsequent recycling of the vesicle membrane. We hypothesized that synaptic vesicle cycling alone, in the absence of neuronal depolarization, could cause Aβ release. To test this premise directly, we developed an acute brain slice model that was amenable to complex pharmacological treatments. Living brain slices (300 μm thick) were produced from three- to four-week old Tg2576 mice and placed into medium at 37 °C for three hours to recover. Media from these slices contain molecules secreted from the brain slice and in many ways are analogous to extracellular brain fluid. Following recovery, slices were incubated with either vehicle or α – *latrotoxin* (LTX), which acts presynaptically to cause synaptic

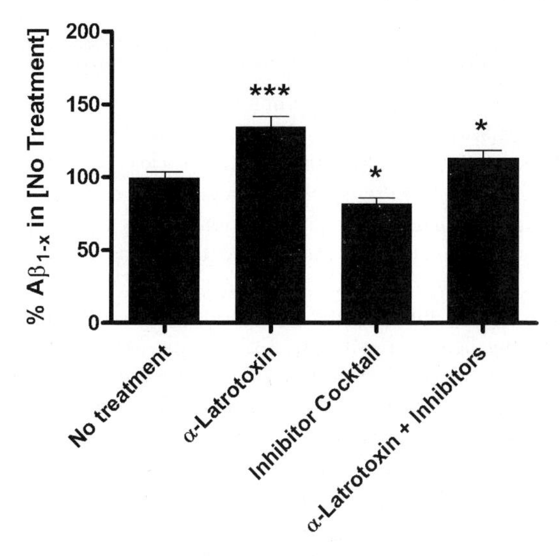

Fig. 3. Synaptic vesicle exocytosis alone is sufficient to elevate ISF Aβ levels in vitro. Acute brain slices were generated from three- to four- week-old Tg2576$^{\pm}$ mice and then maintained at 37 °C in culture media for up to 12 hours. After recovering from slicing for three hours, the slices were treated for two hours with 1) vehicle, 2) α-latrotoxin (0.5 nM), 3) an inhibitor cocktail containing TTX (100 nM), APV (50 μM), and NBQX (10 μM), or 4) a combination of the inhibitor cocktail and α-latrotoxin. Co-treatment with LTX and the inhibitor cocktail caused a 113.3 ± 5.1% increase in Aβ levels within the media compared to untreated slices ($p < 0.05$, $n = 15$ per group). Data expressed as mean ± SEM. Treatment conditions were compared to untreated, control slices using a one-way ANOVA. Reprinted with permission from Neuron

vesicle exocytosis without directly depolarizing the neuron. LTX caused a 135% increase in extracellular Aβ levels (Fig. 3). The neurotransmitter released in response to the toxin, however, included downstream activation of the postsynaptic neuron. To isolate the role of vesicle cycling in the absence of depolarization, we co-administered LTX with a cocktail of activity blockers that included TTX, APV, and NBQX to inhibit sodium channels, NMDA receptors, and AMPA receptors, respectively. The cocktail of inhibitors alone reduced Aβ levels by 20% compared to the untreated condition (Fig. 3), which is comparable to our in vivo results using TTX alone (Fig. 2B). When LTX and the inhibitor cocktail were co-applied to brain slices, extracellular Aβ levels were still significantly elevated (Fig. 3), demonstrating that synaptic vesicle exocytosis alone, in the absence of postsynaptic depolarization, is sufficient to drive Aβ release from neurons.

Proposed Model of Synaptic Activity-Mediated Aβ Release from Neurons

Synaptic activity rapidly and dynamically modulates ISF Aβ levels in vivo. In as little as 30 minutes, seizure activity significantly increased hippocampal ISF Aβ levels whereas a complete blockade of synaptic transmission lowered ISF Aβ levels by 80%. Data from acute brain slices demonstrate that synaptic vesicle cycling is critical for Aβ release. The mechanism that ties synaptic vesicle exocytosis and extracellular Aβ levels together remains unknown, however. Studies employing biochemical and immuno-electron microscopy techniques have not found Aβ within synaptic vesicles (Ikin et al. 1996; Marquez-Sterling et al. 1997). Instead, Aβ exists primarily within the endocytic compartment, suggesting that synaptic vesicle exocytosis does not directly cause Aβ release, though it is likely a closely associated event that is directly responsible for neuronal Aβ release.

Full-length APP is synthesized in the endoplasmic reticulum and is trafficked through the secretory pathway to the plasma membrane. During this journey and at the cell surface, APP is available primarily for cleavage by α-secretase (Sisodia 1992). If full-length APP remains on the cell surface, it undergoes clathrin-mediated endocytosis into the endocytic pathway where it can interact with β- and γ-secretase (Peraus et al., 1997). It appears that most Aβ is generated within the endocytic pathway and that inhibition of endocytosis reduces Aβ production and secretion in vitro (Koo and Squazzo 1994). Once produced in endosomes, Aβ can be shunted to the lysosome for degradation or trafficked back to the plasma membrane, presumably via recycling endosomes, to be secreted from the cell.

When synaptic vesicles release their neurotransmitter, the synaptic vesicle membrane typically fuses with the plasma membrane. Some of that vesicle membrane and the associated proteins are then recycled from the plasma membrane and used to replenish the pool of synaptic vesicles. One mechanism used to remove synaptic membrane from the cell surface is clathrin-mediated endocytosis (Newton et al. 2006). With increased synaptic activity, more synaptic vesicle membrane will be endocytosed from the plasma membrane. We propose that some of this synaptic vesicle membrane recycling impacts the endocytosis of APP, thereby increasing Aβ production and Aβ release (Fig. 4). Within clathrin-coated vesicles, APP co-localizes with several synaptic vesicle markers, namely SV2, synaptotagmin I and synaptophysin (Marquez-Sterling et al. 1997). This finding is consistent with the notion that synaptic vesicle membrane recycling is linked to APP endocytosis. It is possible that activity and APP are only functionally linked and that APP is coincidently endocytosed as synaptic vesicle membrane is recycled.

Synaptic Activity has Several Independent Effects on Aβ Levels

Circumstantial evidence in humans suggests that elevated neuronal activity may be linked with accumulation of Aβ as plaques (Mackenzie and Miller 1994; Buckner et al. 2005). Additionally, experiments in dissociated neuronal culture and in organotypic brain slices demonstrate that modulation of synaptic activity alters extracellular Aβ levels. Studies by our group demonstrate in awake, behaving mice that synaptic transmission can dynamically modulate ISF Aβ levels within minutes. There appear to be at least two mechanisms linking activity and Aβ levels, however. The first mechanism involves postsynaptic receptors, namely M1 acetylcholine receptors and NMDA receptors, which activate signaling pathways resulting in altered processing of APP, thereby affecting Aβ production (Beach et al. 2001; Lesne et al. 2005). It is likely that this mechanism

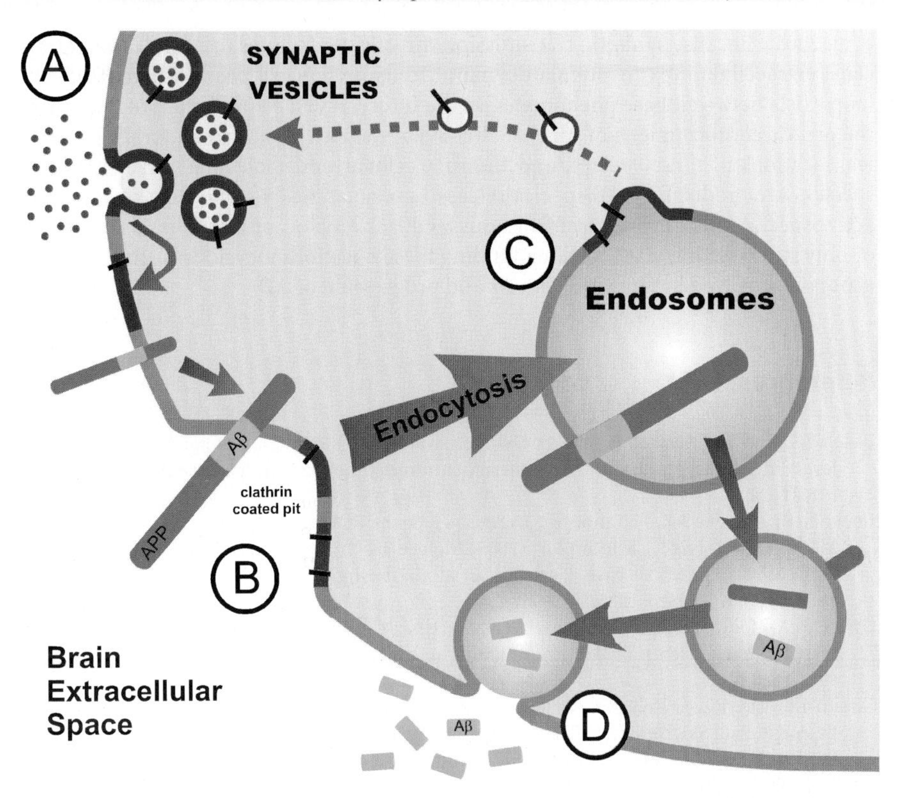

Fig. 4. Proposed model of activity-dependent regulation of Aβ release from neurons. (**A**) Synaptic vesicle exocytosis causes the vesicular membrane to fuse with the plasma membrane. (**B**) Synaptic vesicle membrane is recycled from the cell surface, at least in part via clathrin-mediated endocytosis. Full-length APP resides on the plasma membrane and is endocytosed via clathrin-mediated endocytosis. Increased synaptic activity leads to more vesicle membrane recycling and more coincident APP endocytosis. (**C**) Within the endocytic compartment, APP is cleaved to produce Aβ. (**D**) Aβ is brought back to the cell surface within recycling endosomes and secreted into the brain ISF

appreciably affects Aβ levels over a time course of hours to days. The second mechanism is directly linked to the presynaptic machinery necessary for synaptic transmission. Both increases and decreases in synaptic activity can modulate ISF Aβ levels. This phenomenon appears to be linked directly to synaptic vesicle cycling and occurs on a time scale of minutes. Given that postsynaptic receptor activation can lead to presynaptic vesicle release, these mechanisms may work concurrently within the same neuron.

The Complex Relationship Between Synaptic Transmission and Aβ

A growing literature demonstrates that certain forms of Aβ can inhibit synaptic plasticity (Hartley et al. 1999; Townsend et al. 2006). Additionally, viral-expressed APP depresses synaptic activity in nearby neurons in organotypic brain slices (Kamenetz et al. 2003). Because synaptic transmission directly contributes to Aβ release and because extracellular Aβ can modulate synaptic transmission, it is possible that a feedback loop exists between these phenomena. In the future, it will be important to determine the precise conformations of Aβ that alter synaptic plasticity, the temporal and physical locations of these species, and the root cellular and molecular targets of these species. Are modulatory Aβ species released presynaptically or do they convert into active species following release? Does this feedback loop occur under basal conditions or only in the setting of AD? Understanding these questions may address fundamental processes involved in the pathogenesis and progression of AD.

References

Beach TG, Kuo YM, Schwab C, Walker DG, Roher AE (2001) Reduction of cortical amyloid beta levels in guinea pig brain after systemic administration of physostigmine. Neurosci Lett 310:21–24.

Buckner RL, Snyder AZ, Shannon BJ, LaRossa G, Sachs R, Fotenos AF, Sheline YI, Klunk WE, Mathis CA, Morris JC, Mintun MA (2005) Molecular, structural, and functional characterization of Alzheimer's disease: evidence for a relationship between default activity, amyloid, and memory. J Neurosci 25:7709–7717.

Caccamo A, Oddo S, Billings LM, Green KN, Martinez-Coria H, Fisher A, LaFerla FM (2006) M1 receptors play a central role in modulating AD-like pathology in transgenic mice. Neuron 49:671–682.

Cirrito JR, May PC, O'Dell MA, Taylor JW, Parsadanian M, Cramer JW, Audia JE, Nissen JS, Bales KR, Paul SM, DeMattos RB, Holtzman DM (2003) In vivo assessment of brain interstitial fluid with microdialysis reveals plaque-associated changes in amyloid-beta metabolism and half-life. J Neurosci 23:8844–8853.

Cirrito JR, Yamada KA, Finn MB, Sloviter RS, Bales KR, May PC, Schoepp DD, Paul SM, Mennerick S, Holtzman DM (2005) Synaptic activity regulates interstitial fluid amyloid-beta levels in vivo. Neuron 48:913–922.

Gouras GK, Relkin NR, Sweeney D, Munoz DG, Mackenzie IR, Gandy S (1997) Increased apolipoprotein E epsilon 4 in epilepsy with senile plaques. Ann Neurol 41:402–404.

Gusnard DA, Raichle ME, Raichle ME (2001) Searching for a baseline: functional imaging and the resting human brain. Nature Rev Neurosci 2:685–694.

Hartley DM, Walsh DM, Ye CP, Diehl T, Vasquez S, Vassilev PM, Teplow DB, Selkoe DJ (1999) Protofibrillar intermediates of amyloid beta-protein induce acute electrophysiological changes and progressive neurotoxicity in cortical neurons. J Neurosci 19:8876–8884.

Ikin AF, Annaert WG, Takei K, De Camilli P, Jahn R, Greengard P, Buxbaum JD (1996) Alzheimer amyloid protein precursor is localized in nerve terminal preparations to Rab5-containing vesicular organelles distinct from those implicated in the synaptic vesicle pathway. J Biol Chem 271:31783–31786.

Kamenetz F, Tomita T, Hsieh H, Seabrook G, Borchelt D, Iwatsubo T, Sisodia S, Malinow R (2003) APP Processing and Synaptic Function. Neuron 37:925–937.

Koo EH, Squazzo SL (1994) Evidence that production and release of amyloid beta-protein involves the endocytic pathway. J Biol Chem 269:17386–17389.

Lesne S, Ali C, Gabriel C, Croci N, MacKenzie ET, Glabe CG, Plotkine M, Marchand-Verrecchia C, Vivien D, Buisson A (2005) NMDA receptor activation inhibits alpha-secretase and promotes neuronal amyloid-beta production. J Neurosci 25:9367–9377.

Mackenzie IR, Miller LA (1994) Senile plaques in temporal lobe epilepsy. Acta Neuropathol (Berl) 87:504–510.

Marquez-Sterling NR, Lo AC, Sisodia SS, Koo EH (1997) Trafficking of cell-surface beta-amyloid precursor protein: evidence that a sorting intermediate participates in synaptic vesicle recycling. J Neurosci 17:140–151.

Newton AJ, Kirchhausen T, Murthy VN (2006) Inhibition of dynamin completely blocks compensatory synaptic vesicle endocytosis. Proc Natl Acad Sci USA 103:17955–17960.

Nitsch RM, Deng M, Tennis M, Schoenfeld D, Growdon JH (2000) The selective muscarinic M1 agonist AF102B decreases levels of total Abeta in cerebrospinal fluid of patients with Alzheimer's disease. Ann Neurol 48:913–918.

Peraus GC, Masters CL, Beyreuther K (1997) Late compartments of amyloid precursor protein transport in SY5Y cells are involved in beta-amyloid secretion. J Neurosci 17:7714–7724.

Raichle ME, MacLeod AM, Snyder AZ, Powers WJ, Gusnard DA, Shulman GL (2001) A default mode of brain function. Proc Natl Acad Sci USA 98:676–682.

Sisodia SS (1992) Beta-amyloid precursor protein cleavage by a membrane-bound protease. Proc Natl Acad Sci USA 89:6075–6079.

Sloviter RS, Dichter MA, Rachinsky TL, Dean E, Goodman JH, Sollas AL, Martin DL (1996) Basal expression and induction of glutamate decarboxylase and GABA in excitatory granule cells of the rat and monkey hippocampal dentate gyrus. J Comp Neurol 373:593–618.

Townsend M, Shankar GM, Mehta T, Walsh DM, Selkoe DJ (2006) Effects of secreted oligomers of amyloid beta-protein on hippocampal synaptic plasticity: a potent role for trimers. J Physiol 572:477–492.

Aβ-Induced Toxicity Mediated by Caspase Cleavage of the Amyloid Precursor Protein (APP)

Edward H. Koo[1] and *Dale E. Bredesen*[2]

Summary. Synapse and neuron losses are characteristic features of Alzheimer's disease (AD) and are believed to underlie the cognitive impairments seen in this disorder. Amyloid β-protein (Aβ) is hypothesized to play a pivotal role in initiating AD pathogenesis, but the precise mechanisms of Aβ-induced damage remain unclear. Caspases, a family of proteases best known for their role in programmed cell death, may play a role in neurodegeneration. This review will focus on recent findings implicating a role for caspase cleavage of the amyloid precursor protein (APP) in synaptic and neuronal injury in AD and how this pathway may be initiated by the interaction of Aβ with APP.

Introduction

The pathological hallmark of AD, the most common age-related neurodegenerative disorder, is the presence of neurofibrillary tangles (NFTs) and extracellular deposits of amyloid β-protein (Aβ) in senile plaques in cerebral cortex. Although these brain lesions are seen in aged, non-demented individuals, the formation of amyloid in brain is increasingly viewed as the earliest histopathologic event in the pathogenesis of AD. This view is at the heart of the decade-old "amyloid cascade hypothesis," which states that neuronal dysfunction and death, synapse loss, neurofibrillary degeneration, microglial activation and the full manifestation of Alzheimer pathology are initiated by the accumulation of amyloid deposition in the brain (Hardy and Selkoe 2002). In the past several years, however, it has become more evident that amyloid in its pre-fibrillar form may be more damaging to the brain. This finding has led to the concept of soluble oligomeric Aβ as perhaps being the critical amyloid species that plays the causal role in initiating this cascade (Haass and Selkoe 2007). This short article will review some of the recent findings concerning a pathway of Aβ-induced toxicity that is APP-dependent. Specifically, the role of caspase cleavage of APP in neurotoxicity and the implications of these findings for synaptic damage in AD will be discussed.

Synaptic Pathology in AD

Although both senile plaques and NFTs define the pathology of AD, NFTs correlate better than senile plaques with premortem cognitive decline. Neuronal loss, although

[1] Department of Neurosciences, University of California, San Diego, La Jolla, CA 92037, email: edkoo@ucsd.edu

[2] Buck Institute for Age Research, Novato, CA 94945 and Department of Neurology, University of California, San Francisco, San Francisco, CA 94143

Selkoe et al.
Synaptic Plasticity and the Mechanism of Alzheimer's Disease
© Springer-Verlag Berlin Heidelberg 2008

a characteristic of all neurodegenerative disorders, is not well correlated with disease progression. However, the best correlate of dementia appears to be the degree of synapse loss in brain (Terry et al. 1991). Meticulous ultrastructural studies have documented the loss of synapses in AD brain (Scheff et al. 1990). Importantly, the loss of between 25–35% of synapses was documented from cortical biopsies, rather than from postmortem examination, in AD cases of relatively short duration (DeKosky and Scheff 1990). Using synaptophysin immunoreactivity as a measure of synapse numbers, the density of neocortical synaptophysin puncta in postmortem brain tissue was shown to be the best correlate to premortem cognition, substantially better than either the number of NFTs, senile plaque density, or neuronal count. Indeed, regression analysis suggested that symptomatic dementia appeared when there was about 40% loss of neocortical synapses as compared to unaffected controls (Terry et al. 1991). Synapse loss is also seen in presymptomatic individuals with AD lesions (preclinical AD) and in the early stages of AD, when subtle impairment of memory is first manifested as mild cognitive impairment (MCI). Quantitative analysis of the neocortical density of neurons and synapses indicated that synapse loss exceeded neuron loss, and ultrastructural counts of synapse numbers showed a reduction in dentate gyrus in MCI individuals that was intermediate between that in controls and AD individuals (Scheff et al. 2005). Thus, it appears that synapses are damaged and lost when the neurons are still viable (Selkoe 2002; Coleman et al. 2004). Consequently, there is growing consensus in the field concerning the importance of synaptic health in the development of disease phenotype.

Despite the abundance of evidence implicating synapses as targets of Aβ-induced toxicity, little is known about the mechanistic pathways leading downstream from Aβ to synaptic dysfunction or damage. Recent findings suggest that the soluble pool of Aß peptides, rather than the highly aggregated and fibrillar forms of Aβ, are more pathologically relevant. For example, premortem cognitive status is best correlated with the levels of the soluble pool of Aβ peptides in postmortem brain (Lue et al. 1999). Moreover, in transgenic mice that over-express the human APP and develop age-related Aβ deposits, the reduction in synaptophysin immunoreactivity, electrophysiological abnormalities, hippocampal atrophy, and behavioral changes are seen well before any evidence of plaque formation and neuronal death but correlate with a rise in Aβ levels (Mucke et al. 2000; Redwine et al. 2003; Lanz et al. 2003; Wu et al. 2004; Buttini et al. 2005). Furthermore, deficits in long-term potentiation (LTP) and synapse loss can be reversed in APP transgenic mice upon removal of amyloid by passive immunization with Aβ-specific antibodies, thus lending further weight to the idea that Aβ is synaptotoxic (Lombardo et al. 2003; Buttini et al. 2005). And even after plaque formation, axonal and dendritic spine abnormalities, especially those around plaque deposits, are rapidly reversed when the Aβ is removed (Brendza et al. 2005).

APP in Alzheimer's Disease Pathogenesis

The APP gene is located on chromosome 21. This chromosomal localization immediately tied APP to trisomy 21 (Down's syndrome) individuals who all develop Alzheimer's pathology indistinguishable from non-trisomic individuals by the fourth decade of life. Indeed, triplication of this gene appears to be sufficient for Down's syndrome-affected individuals to develop AD pathology (Prasher et al. 1998). Importantly, triplication of only the APP locus has now been shown to be causative of one form of autosomal dominant AD without trisomy 21 (Rovelet-Lecrux et al. 2006). In mice with segmental trisomy 16, a rodent model of human trisomy 21, a number of alterations in brain, including morphological changes that are reminiscent of those seen in AD, have been shown to be due to triplication of the APP gene (Cataldo et al. 2003; Salehi et al. 2006). Together, these findings provide the best evidence for a central role of APP in AD pathogenesis. Whether it is only a mild 50% increase in APP or an increase in Aβ, or both, that is sufficient to cause AD in these genetic forms of AD, and potentially in sporadic cases as well, remains to be clarified.

APP is a member of a family of three proteins, but the other two members, APLP1 and APLP2, do not contain the Aβ sequence. All are type I membrane proteins with a large extracellular domain and a short cytoplasmic region. Alternative splicing generates multiple isoforms, with the 695 amino acid residue isoform (APP695) being the predominant and neuron-specific species in the brain. APP undergoes constitutive cleavage by α-secretases (TACE and ADAM family of proteases), releasing the large, soluble N-terminal derivative, APPsα. This cleavage event takes place both intracellularly and on the cell surface. Significantly, the cleavage site occurs within the Aβ domain and thus generation of APPsα precludes the formation of full-length Aβ peptides (Sisodia et al. 1990). APP molecules can alternatively be cleaved by secretase (BACE1) to generate the N-terminus of Aβ (Vassar et al. 1999). A final cleavage within the transmembrane region generates the C-terminus and releases Aβ from APP. This intramembrane cleavage event is designated γ-secretase activity and requires a complex of at least four proteins: presenilin-1 (PS1) or presenilin-2 (PS2) as well as nicastrin, APH-1, and PEN-2 (Kopan and Ilagan 2004). This presenilin complex also mediates the cleavage of an increasing number of membrane substrates, including Notch receptors in which γ-secretase cleavage is required to release the cytosolic domain for normal signaling activities. Following γ-secretase or, more specifically, ε-secretase cleavage in the case of APP, the cytosolic tail of APP, called APP intracellular domain (AID or AICD), is released from the membrane. By analogy to the Notch receptor, it has been proposed that AICD may also have signaling properties, however AICD is normally rapidly degraded and physiologic data in support of this idea are still wanting.

APP is a Caspase Substrate

In addition to the α-, β-, and γ-secretase cleavages that release the APP ectodomain (APPs) and various Aβ species, APP can be cleaved by caspases. Although APP is cleaved in multiple sites by caspases in vitro (Gervais et al. 1999), there appears to be a primary cleavage site in the cytosolic domain after the aspartate residue at position 664 (using APP 695 numbering), such that mutation of this aspartate residue to alanine (APP D664A) abrogates this proteolytic event (Weidemann et al. 1999; Pellegrini et al. 1999; Lu et al. 2000; Fig. 1). APP cleavage at this position by caspases or caspase-like proteases has been detected in vivo, indicating that this event occurs physiologically (Gervais et al. 1999; Lu et al. 2000). The VEVD motif is a canonical substrate for caspase-6 cleavage, but multiple caspases can cleave APP when tested in vitro or in cell culture. However, it is unclear which caspase(s) is primarily responsible for cleavage in the in vivo setting.

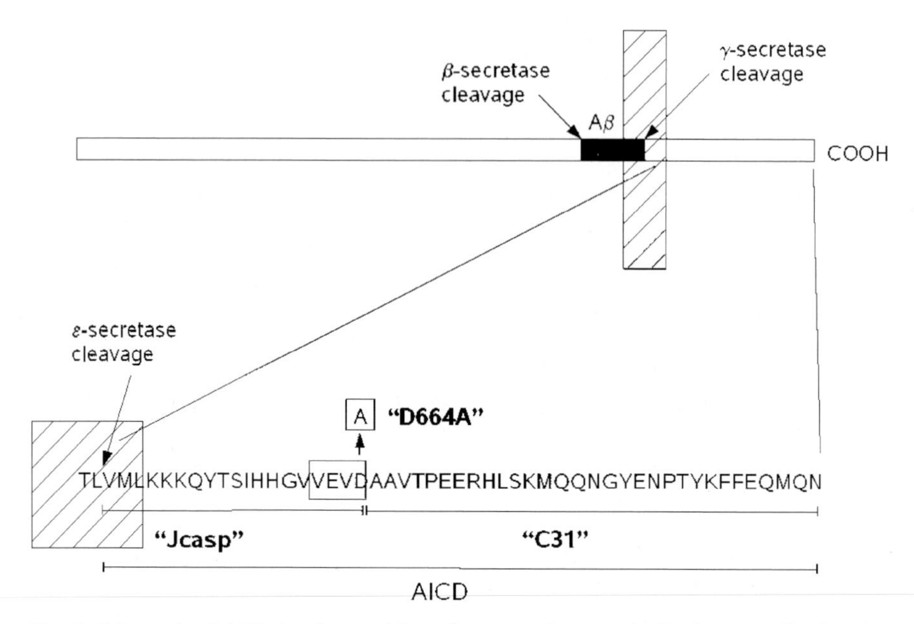

Fig. 1. Schematic of APP showing position of caspase cleavage site in the cytosolic domain. Cleavage at position D664 would release a C-terminal peptide of 31 amino acid residues dubbed C31. In addition, another peptide, Jcasp, can be released after ε-secretase cleavage that normally liberates the APP intracellular domain, AICD. The locations of Aβ peptide, D664A mutation, and β-, ε-, and γ-secretase sites are shown

Caspases and Apoptosis in AD

The name caspase derives from cysteine-dependent aspartate-specific protease. More than a dozen caspases have been identified, and these proteases have stringent specificity for cleaving substrates containing aspartate residue. In addition, catalysis is governed by a critical conserved cysteine side chain of the enzyme (Timmer and Salvesen 2007). Caspases participate in programmed cell death or in the apoptotic pathway and are separated into functional groups (Kumar 2007). Initiator caspases (caspase-2, -8, -9, -10) are first activated by an apoptotic stimulus and then activate the effector caspases (caspase-3, -6, and -7). The latter, once activated, cleave a number of cellular proteins and, in so doing, lead to cell death. This pathway is activated through two broad mechanisms. The extrinsic pathway is activated by specific ligands to cell surface receptors, such as the TNF receptor family, to recruit and activate caspase-8, which in turn activates the effector caspases. The intrinsic pathway centers in the mitochondria, whereby release of cytochrome c after mitochondrial membrane depolarization leads to formation of the apoptosome complex that activates caspase-9 and, in turn, activates the effector caspases. Caspases cleave not only the cellular proteins that activate the death signaling pathway but also a number of cellular substrates that are unrelated to cell death. Thus, the detection of cleavage of these substrates can serve as a marker of caspase activation in situ.

Whether classical apoptosis, defined morphologically by cell shrinkage, chromatin condensation, and membrane blebbing with preservation of organelle structure, is a major mechanism responsible for neuronal death in AD is unknown. Initial enthusiasm for apoptosis in AD has been dampened by the lack of morphological evidence of apoptosis in the brain (LeBlanc 2005; Jellinger 2006). Furthermore, although Aβ has been shown to cause apoptosis in a variety of cells in culture, most studies have used high concentrations of peptide that may not be physiologic. In addition, positivity in neurons to end-specific labeling of fragmented DNA by the TUNEL method was interpreted to imply DNA fragmentation that occurs during apoptosis. However, these findings have been difficult to confirm and some of the positive findings have been reinterpreted to represent non-specific DNA damage unrelated to apoptosis (Stadelmann et al, 1998). Nevertheless, using antibodies specific to the active subunits of caspases, studies have reported the activation of caspase-8 and -9 in brains of AD individuals (Rohn et al. 2001a, 2002). However, there is no consistent evidence that caspase-3, an effector caspase that is downstream of caspase-8 and -9, is activated, as would be expected (LeBlanc 2005). Interestingly, an antibody specific to activated caspase-6 immunostained dystrophic neurites, neuropil threads, and NFTs in the brains of AD individuals (Guo et al. 2004). This latter observation is consistent with the detection of caspase-cleaved tau in neurons, especially those bearing NFTs (Rissman et al. 2004; Albrecht et al. 2007). In addition, caspase-cleaved actin and fodrin are detected in AD brains, thus providing further, albeit indirect, evidence of caspase activation (Yang et al. 1998; Rohn et al. 2001b). In sum, although classical apoptosis may not be a generalized pathway of neuronal cell death in AD, there is nevertheless good evidence that caspases are activated to some degree in AD and therefore could play a role in AD pathogenesis.

Caspase-Mediated APP Cleavage and Neuronal Injury

The interest in APP cleavage by caspases is derived from a number of observations. As mentioned above, APP has been shown by a number of laboratories to be a substrate for caspase cleavage. Importantly, we first reported that cleavage of APP at position D664 releases a novel cytotoxic peptide called C31, for a C-terminal APP peptide of 31 amino acid residues long (Lu et al. 2000; Fig. 1). In vitro experiments showed that C31 causes cell death in a variety of cultured cells, either following cDNA expression or direct transduction with synthetic C31 peptide (Lu et al. 2000; Galvan et al. 2002). Interestingly, another APP-derived peptide, called Jcasp, has also been reported to be cytotoxic (Bertrand et al. 2001). This peptide is situated between the γ-secretase and caspase cleavage sites and presumably would be generated after AICD is re-cleaved by caspases or caspase-cleaved APP is re-cleaved by γ-secretase (Fig. 1). Thus, it is possible that two APP-derived cytotoxic peptides are produced following cleavage at position D664.

To test this model of neuronal toxicity in vivo, new lines of APP transgenic mice were generated in which this intracytoplasmic caspase is mutated from aspartate to alanine (D664) to prevent cleavage (Galvan et al. 2006). The construct is designed to test whether this cleavage site contributes to the synapse loss and dysfunction that are concomitant with amyloid pathology in the brain. Consequently, the construct was modified from that used by the Mucke lab to generate the original J9 and J20 lines of APP of transgenic mice (PD-APP) that exhibit abundant Aß deposits in the brain, by expressing the double Swedish (K595N/M596L) and Indiana APP (V642F) mutations (Mucke et al. 2000). The D664A mutation was added to the original APP familial AD mutations to generate the APP (Swedish/Indiana)-D664A transgenic mice, herein designated APP D664A. Multiple APP D664A transgenic mouse lines were generated with different levels of APP expression and the analyses were concentrated on two lines, B21 and B254. APP expression in the B21 mice was slightly lower than the J20 PD-APP mice but higher in the B254 line (i.e., B21 < J20 < B254). Consistent with this difference in APP expression, B21 animals showed amyloid deposits in the brain that were similar to those in J20 animals at 8–10 months, whereas the B254 mice developed Aβ deposits in the brain by six months of age. These findings represent a confirmation of the cell-based studies showing that Aβ production is not perturbed by the D664A mutation. Next, the Bredesen laboratory evaluated synaptic densities based on synaptophysin staining in control, J9, J20, B21, and B254 mice. In blinded studies, they confirmed that hippocampal synaptic densities are reduced by 30–50% in the J9 and J20 PD-APP mice, as reported by Mucke and colleagues in the original description of these transgenic mouse lines. However, the APP D664A mice did not show any loss in synaptophysin immunoreactivity as compared to littermate controls, indicating preservation of synapse numbers. Furthermore, other associated morphological changes were also different between the APP D664A mice and the PD-APP mice. Specifically, there was an absence of dentate gyral atrophy, reduced astrogliosis, and a lack of behavioral impairments in learning and memory in APP D664A mice, as would be expected of "conventional" APP transgenic mice.

Because synaptophysin immunoreactivity is only a surrogate measure of synaptic density, direct assessment of synaptic function is important. Consequently, a second study was carried out in APP D664A transgenic mice to ascertain synaptic function. (Saganich et al. 2006). In PD-APP J20 mice, there was an approximately 35% reduction in basal synaptic transmission at the Schaffer collateral to CA1 synapse in acute hippocampal slices from three- to six-month-old animals. In contrast, APP D664A B21 line transgenic mice at the same ages showed no change in basal synaptic function as compared to littermate non-transgenic controls. The higher expressing APP D664A B254 line showed only a mild but non-significant 12% reduction in synaptic function. Importantly, five- to six-month-old B254 line mice also did not show any deficits, at a time when amyloid deposits were already detected. Therefore, even in the presence of high Aβ levels and early plaque formation, there was still no evidence of a significant drop in synaptic transmission in the hippocampus in mice where APP has been mutated to prevent caspase cleavage. Finally, synaptic plasticity was assessed by measuring LTP. Similar to the basal synaptic transmission results, LTP was also normal in both B21 and B254 lines of APP D664A transgenic mice but was depressed in the PD-APP transgenic mice. Taken together, the results suggested that the D664A mutation protected APP transgenic mice from neurodegeneration with respect to synaptic injury when compared to APP transgenic mice without this mutation, while maintaining comparable Aß levels and amyloid deposits in brain.

Several questions are not yet answered in the APP D664A mice. First, it is unclear whether the lack of synaptic damage in APP D664A mice is due solely to the lack of

C31 generation and potentially Jcasp fragment production. For example, the D664A mutation could have compromised other protein–protein interactions that contribute to synaptic toxicity. Second, it is unclear which caspase is cleaving APP to form C31 (although previous work in cell culture demonstrated the interaction of caspase-8 with APP; Lu et al. 2000). Third, the context in which caspase is activated to initiate this pathway is not yet known. We have proposed that Aβ could be one of the triggers (discussed below), but in vivo confirmation is lacking so far. Fourth, even if C31 is the primary trigger, the precise mechanism by which C31 is cytotoxic has not been established. Fifth, the APP D664A mice are newly generated and thus do not present a formal reversal of deficits seen in control APP transgenic mice without mutation at this position. Central to our argument is that these deficits are predicted to occur in transgenic mice that over-express human APP and that show age-associated amyloid deposits. However, not all APP transgenic mouse lines show comparable profiles of abnormalities. For example, most reports showed reductions in basal synaptic transmission, but these reductions are by no means uniform (Moechars et al. 1999; Hsia et al. 1999; Giacchino et al. 2000; Fitzjohn et al. 2001). One study reported no alterations but this could be due to differences in the methodologies that were used (Chapman et al. 1999). Similarly, LTP was not uniformly depressed in all APP transgenic mice in which it was examined (Saganich et al. 2006). Finally, as we have proposed, this is likely to be but one of the pathways that contribute to synaptic damage in AD. Other pathways are certainly involved in Aβ-induced toxicity, some of which are discussed in other chapters of this volume.

Parallels to Huntington's Disease

The above findings show striking similarities to a recent report in mice that express expanded polyglutamine (polyQ) repeats in the huntingtin (Htt) gene and that show features associated with Huntington's disease (HD) in humans, including movement and behavioral deficits, neurodegeneration in the striatum, and sensitivity to excitotoxic damage. HD is a neurodegenerative disorder resulting from expanded polyQ repeats in the Htt gene. Interestingly, Htt is also a substrate for caspase cleavage, and N-terminal Htt fragments, which can be released after caspase cleavage, accumulate in neuronal nuclei. Strikingly, mice with expanded polyQ expansion and resistant to caspase-6 but not to caspase-3 cleavage failed to develop any of the HD-like phenotypes (Graham et al. 2006). Specifically, these mice did not show any motor abnormalities, striatal degeneration, accumulation of Htt fragments in neuronal nuclei, or enhanced excitotoxicity. And, as in the case of APP, caspase cleavage of Htt has been shown to generate a toxic fragment that increased the susceptibility of neurons to excitotoxicity in cultured cells (Wellington et al., 2000). Lastly, although the precise mechanism of toxicity due to caspase-cleaved Htt is still unclear, the parallels to APP are nevertheless striking.

Aβ and APP Interactions in APP Caspase-Mediated Cleavage

At first blush, the findings from the APP D664A mice appear to challenge the role of Aβ in synaptic toxicity in AD and in APP transgenic mice. On the contrary, the results are entirely consistent with a model emanating from cell culture studies implicating both Aβ and APP in a common mechanistic pathway underlying cellular dysfunction. Specifically, it has been shown that loss of APP attenuates Aβ toxicity (Lorenzo et al. 2000). Second, Aβ treatment of cells increases cleavage of APP at this position. Third, both Aβ- and C31-induced cytotoxicity activate a number of caspases and cell death from both insults can be blocked by a similar profile of caspase inhibitors (Lu et al. 2003b). Fourth, we have proposed that one mechanism of Aβ-induced toxicity is through the binding of Aβ to APP causing APP to form a complex, an event that may trigger cell death (Lu et al. 2003a). Subsequently, it has been shown that Aβ binds to the cognate Aβ region in the extracellular portion of APP to facilitate APP dimerization. (Shaked et al. 2006). In these studies, the dimerization of APP has been shown to recruit caspase-8 to the complex, leading to subsequent cleavage of APP at D664. Furthermore, formation of the APP complex appears to be the critical step, as forced dimerization of APP by artificial cross-linking of APP results in cytotoxicity even in the absence of Aβ (Rohn et al. 2000; Lu et al. 2003a). However, introducing the D664A mutation in this context attenuates cell death caused by artificial APP dimerization. Fifth, the D664A mutation substantially reduces Aβ-induced cell death. Finally, antibodies specific to the C-terminus of APP generated after caspase cleavage showed that these APP species truncated at D664 are increased in the brains of AD individuals and in APP transgenic mice (Su et al. 2002; Ayala-Grosso et al. 2002; Zhao et al. 2003). These findings therefore suggest a model whereby Aβ toxicity is, at least in part, APP dependent, and this pathway is specifically mediated by cleavage of APP at D664. Taken together, the inference from these studies is that the D664A mutation leads to a loss of caspase cleavage at that position, reducing the generation of C31 and potentially Jcasp as well and, in so doing, diminishes synaptic injury and cell death.

Conclusion

Although our current data fit into this working model, the precise mechanism of neurotoxicity in this Aβ-APP pathway remains unknown. Nevertheless, if prevention of D6664A cleavage is indeed protective for synaptic function in the presence of Aβ, then uncovering the mechanism by which this cleavage contributes to synaptic dysfunction and neuronal injury will be important, because insights into this pathway have implications for both understanding synaptic damage in AD and developing drugs to treat AD by targeting the synapse.

References

Albrecht S, Bourdeau M, Bennett D, Mufson EJ, Bhattacharjee M, LeBlanc AC (2007) Activation of caspase-6 in aging and mild cognitive impairment. Am J Pathol 170:1200–1209.

Ayala-Grosso C, Ng G, Roy S, Robertson GS (2002) Caspase-cleaved amyloid precursor protein in Alzheimer's disease. Brain Pathol 12:430–441.

Bertrand E, Brouillet E, Caille I, Cole GM, Prochiantz A, Allinquant B (2001) A short cytoplasmic domain of the amyloid precursor protein induces apoptosis *in vitro* and *in vivo*. Mol Cell Neurosci 18:503–511.

Brendza RP, Bacskai BJ, Cirrito JR, Simmons KA, Skoch JM, Klunk WE, Mathis CA, Bales KR, Paul SM, Hyman BT, Holtzman DM (2005) Anti-Aβ antibody treatment promotes the rapid recovery of amyloid-associated neuritic dystrophy in PDAPP transgenic mice. J Clin. Invest 115:428–433.

Buttini M, Masliah E, Barbour R, Grajeda H, Motter R, Johnson-Wood K, Khan K, Seubert P, Freedman S, Schenk D, Games D (2005) β-amyloid immunotherapy prevents synaptic degeneration in a mouse model of Alzheimer's disease. J Neurosci 25:9096–9101.

Cataldo AM, Petanceska S, Peterhoff CM, Terio NB, Epstein CJ, Villar A, Carlson EJ, Staufenbiel M, Nixon RA (2003) App gene dosage modulates endosomal abnormalities of Alzheimer's disease in a segmental trisomy 16 mouse model of down syndrome. J Neurosci 23:6788–6792.

Chapman PF, White GL, Jones MW, Cooper-Blacketer D, Marshall VJ, Irizarry M, Younkin L, Good MA, Bliss TV, Hyman BT, Younkin SG, Hsiao KK (1999) Impaired synaptic plasticity and learning in aged amyloid precursor protein transgenic mice. Nature Neurosci 2:271–276.

Coleman PD, Federoff H, Kurlan R (2004) A focus on the synapse for neuroprotection in Alzheimer's disease and other dementias. Neurology 63:1155–1162.

DeKosky ST, Scheff SW (1990) Synapse loss in frontal cortex biopsies in Alzheimer's disease: correlation with cognitive severity. Ann Neurol 27:457–464.

Fitzjohn SM, Morton RA, Kuenzi F, Rosahl TW, Shearman M, Lewis H, Smith D, Reynolds DS, Davies CH, Collingridge GL, Seabrook GR (2001) Age-related impairment of synaptic transmission but normal long-term potentiation in transgenic mice that overexpress the human APP695SWE mutant form of amyloid precursor protein. J Neurosci. 21:4691–4698.

Galvan V, Chen S, Lu D, Logvinova A, Goldsmith P, Koo EH, Bredesen DE (2002) Caspase cleavage of members of the amyloid precursor family of proteins. J Neurochem 82:283–294.

Galvan V, Gorostiza OF, Banwait S, Ataie M, Logvinova AV, Sitaraman S, Carlson E, Sagi SA, Chevallier N, Jin K, Greenberg DA, Bredesen DE (2006) Reversal of Alzheimer's-like pathology and behavior in human APP transgenic mice by mutation of Asp664. Proc Natl Acad Sci USA 103:7130–135.

Gervais FG, Xu D, Robertson GS, Vaillancourt JP, Zhu Y, Huang J, LeBlanc A, Smith DW, Rigby M, Shearman MS, Clarke EE, Zheng H, Van der Ploeg HT, Ruffolo SC, Thornberry NA, Xanthoudakis S, Zamboni RJ, Roy S, Nicholson DW (1999) Involvement of Caspases in proteolytic cleavage of Alzheimer's amyloid-β precursor protein and amyloidogenic Aβ peptide formation. Cell 97:395–406.

Giacchino J, Criado JR, Games D, Henriksen S (2000) In vivo synaptic transmission in young and aged amyloid precursor protein transgenic mice. Brain Res 876:185–190.

Graham RK, Deng Y, Slow EJ, Haigh B, Bissada N, Lu G, Pearson J, Shehadeh J, Bertram L, Murphy Z, Warby SC, Doty CN, Roy S, Wellington CL, Leavitt BR, Raymond LA, Nicholson DW, Hayden MR (2006) Cleavage at the caspase-6 site is required for neuronal dysfunction and degeneration due to mutant huntingtin. Cell 125:1179–1191.

Guo H, Albrecht S, Bourdeau M, Petzke T, Bergeron C, LeBlanc AC. (2004) Active caspase-6 and caspase-6-cleaved tau in neuropil threads, neuritic plaques, and neurofibrillary tangles of Alzheimer's disease. Am J Pathol 165:523–531.

Haass C, Selkoe DJ (2007) Soluble protein oligomers in neurodegeneration: lessons from the Alzheimer's amyloid β-peptide. Nature Rev Mol Cell Biol 8:101–112.

Hardy J, Selkoe DJ (2002) The amyloid hypothesis of Alzheimer's disease: progress and problems on the road to therapeutics. Science 297:353–356.

Hsia AY, Masliah E, McConlogue L, Yu GQ, Tatsuno G, Hu K, Kholodenko D, Malenka RC, Nicoll RA, Mucke L (1999) Plaque-independent disruption of neural circuits in Alzheimer's disease mouse models. Proc Natl Acad Sci USA 96:3228–3233.

Jellinger KA (2006) Challenges in neuronal apoptosis. Curr Alzheimer Res 3:377–391.

Kopan R, Ilagan MX (2004) γ-secretase: proteosome of the membrane? Nature Rev Mol Cell Biol 5:499–504.

Kumar S (2007) Caspase function in programmed cell death. Cell Death Differ 14:32–43.

Lanz TA, Carter DB, Merchant KM (2003) Dendritic spine loss in the hippocampus of young PDAPP and Tg2576 mice and its prevention by the ApoE2 genotype. Neurobiol Dis 13:246–253.

LeBlanc AC (2005) The role of apoptotic pathways in Alzheimer's disease neurodegeneration and cell death. Curr Alzheimer Res 2:389–402.

Lombardo JA, Stern EA, McLellan ME, Kajdasz ST, Hickey GA, Bacskai BJ, Hyman BT (2003) Amyloid-β antibody treatment leads to rapid normalization of plaque-induced neuritic alterations. J Neurosci 23:10879–10883.

Lorenzo A, Yuan M, Zhang Z, Paganetti PA, Sturchler-Pierrat C, Staufenbiel M, Mautino J, Vigo FS, Sommer B, Yankner BA (2000) Amyloid β interacts with the amyloid precursor protein: a potential toxic mechanism in Alzheimer's disease. Nature Neurosci 3:460–464.

Lu DC, Rabizadeh S, Chandra S, Shayya RF, Ellerby LM, Ye X, Salvesen GS, Koo EH, Bredesen DE (2000) A second cytotoxic proteolytic peptide derived from amyloid β-protein precursor. Nature Med 6:397–404.

Lu DC, Shaked GM, Masliah E, Bredesen DE, Koo EH (2003a) Amyloid beta protein toxicity mediated by the formation of amyloid-β protein precursor complexes. Ann Neurol 54:781–789.

Lu DC, Soriano S, Bredesen DE, Koo EH (2003b) Caspase cleavage of the amyloid precursor protein modulates amyloid beta-protein toxicity. J Neurochem 87:733–741.

Lue L-F, Kuo Y-M, Roher AE, Brachova L, Shen Y, Sue L, Rydel RE, Rogers J (1999) Soluble amyloid β peptide concentration as a predictor of synaptic change in Alzheimer's disease. Am J Pathol 155:853–862.

Moechars D, Dewachter I, Lorent K, Reverse D, Baekelandt V, Naidu A, Tesseur I, Spittaels K, Haute CV, Checler F, Godaux E, Cordell B, Van Leuven F (1999) Early phenotypic changes in transgenic mice that overexpress different mutants of amyloid precursor protein in brain. J Biol Chem 274:6483–6492.

Mucke L, Masliah E, Yu GQ, Mallory M, Rockenstein EM, Tatsuno G, Hu K, Kholodenko D, Johnson-Wood K, McConlogue L (2000) High-level neuronal expression of Aβ1-42 in wild-type human amyloid protein precursor transgenic mice: synaptotoxicity without plaque formation. J Neurosci 20:4050–4058.

Pellegrini L, Passer BJ, Tabaton M, Ganjei JK, D'Adamio L (1999) Alternative, non-secretase processing of Alzheimer's β-amyloid precursor protein during apoptosis by Caspase-6 and -8. J.Biol.Chem. 274:21011–6.

Prasher VP, Farrer MJ, Kessling AM, Fisher EMC, West RJ, Barber PC, Butler AC. (1998) Molecular mapping of Alzheimer-type dementia in Down's syndrome. Ann Neurol 43:380–383.

Redwine JM, Kosofsky B, Jacobs RE, Games D, Reilly JF, Morrison JH, Young WG, Bloom FE (2003) Dentate gyrus volume is reduced before onset of plaque formation in PDAPP mice: a magnetic resonance microscopy and stereologic analysis. Proc Natl Acad Sci USA 100:1381–1386.

Rissman RA, Poon WW, Blurton-Jones M, Oddo S, Torp R, Vitek MP, LaFerla FM, Rohn TT, Cotman CW (2004) Caspase-cleavage of tau is an early event in Alzheimer disease tangle pathology. J Clin Invest 114:121–130.

Rohn TT, Head E, Nesse WH, Cotman CW, Cribbs DH (2001a) Activation of caspase-8 in the Alzheimer's disease brain. Neurobiol Dis 8:1006–1016.

Rohn TT, Head E, Su JH, Anderson AJ, Bahr BA, Cotman CW, Cribbs DH (2001b) Correlation between caspase activation and neurofibrillary tangle formation in Alzheimer's disease. Am J Pathol 158:189–198.

Rohn TT, Ivins KJ, Bahr BA, Cotman CW, Cribbs DH (2000) A monoclonal antibody to amyloid precursor protein induces neuronal apoptosis. J Neurochem 74:2331–2342.

Rohn TT, Rissman RA, Davis MC, Kim YE, Cotman CW, Head E (2002) Caspase-9 activation and caspase cleavage of tau in the Alzheimer's disease brain. Neurobiol Dis 11:341–354.

Rovelet-Lecrux A, Hannequin D, Raux G, Le MN, Laquerriere A, Vital A, Dumanchin C, Feuillette S, Brice A, Vercelletto M, Dubas F, Frebourg T, Campion D (2006) APP locus duplication causes autosomal dominant early-onset Alzheimer disease with cerebral amyloid angiopathy. Nature Genet 38:24–26.

Saganich MJ, Schroeder BE, Galvan V, Bredesen DE, Koo EH, Heinemann SF (2006) Deficits in synaptic transmission and learning in amyloid precursor protein (APP) transgenic mice require C-terminal cleavage of APP. J Neurosci 26:13428–13436.

Salehi A, Delcroix JD, Belichenko PV, Zhan K, Wu C, Valletta JS, Takimoto-Kimura R, Kleschevnikov AM, Sambamurti K, Chung PP, Xia W, Villar A, Campbell WA, Kulnane LS, Nixon RA, Lamb BT, Epstein CJ, Stokin GB, Goldstein LS, Mobley WC (2006) Increased App expression in a mouse model of Down's syndrome disrupts NGF transport and causes cholinergic neuron degeneration. Neuron 51:29–42.

Scheff SW, DeKosky ST, Price DA (1990) Quantitative assessment of cortical synaptic density in Alzheimer's disease. Neurobiol Aging 11:29–37.

Scheff SW, Price DA, Schmitt FA, Mufson EJ (2006) Hippocampal synaptic loss in early Alzheimer's disease and mild cognitive impairment. Neurobiol Aging. 27:1372–1384.

Selkoe DJ (2002) Alzheimer's disease is a synaptic failure. Science 298:789–791.

Shaked GM, Kummer MP, Lu DC, Galvan V, Bredesen DE, Koo EH (2006) Aβ induces cell death by direct interaction with its cognate extracellular domain on APP (APP 597–624). FASEB J 20:1254–1256.

Sisodia SS, Koo EH, Beyreuther K, Unterbeck A, Price DL (1990) Evidence that β-amyloid protein in Alzheimer's disease is not derived by normal processing. Science 248:492–495.

Stadelmann C, Bruck W, Bancher C, Jellinger K, Lassmann H (1998) Alzheimer disease: DNA fragmentation indicates increased neuronal vulnerability, but not apoptosis. J Neuropathol Exp Neurol 57:456-464.

Su JH, Kesslak JP, Head E, Cotman CW (2002) Caspase-cleaved amyloid precursor protein and activated caspase-3 are co-localized in the granules of granulovacuolar degeneration in Alzheimer's disease and Down's syndrome brain. Acta Neuropathol (Berl) 104:1–6.

Terry RD, Masliah E, Salmon DP, Butters N, DeTeresa R, Hill R, Hansen LA, Katzman R (1991) Physical basis of cognitive alterations in Alzheimer's disease: synapse loss is the major correlate of cognitive impairment. Ann Neurol 30:572–580.

Timmer JC, Salvesen GS (2007) Caspase substrates. Cell Death Differ 14:66–72.

Vassar R, Bennett BD, Babu-Khan S, Kahn S, Mendiaz EA, Denis P, Teplow DB, Ross S, Amarante P, Loeloff R, Luo Y, Fisher S, Fuller J, Edenson S, Lile J, Jarosinski MA, Biere AL, Curran E, Burgess T, Louis JC, Collins F, Treanor J, Rogers G, Citron M (1999) β-secretase cleavage of Alzheimer's amyloid precursor protein by the transmembrane aspartic protease BACE. Science 286:735–741.

Weidemann A, Paliga K, Durrwang U, Reinhard JFJ, Schuckert O, Evin G, Masters CL (1999) Proteolytic processing of the Alzheimer's Disease amyloid precursor protein within its cytoplasmic domain by caspase-like proteases. J Biol Chem 274:5823–5829.

Wellington CL, Singaraja R, Ellerby L, Savill J, Roy S, Leavitt B, Cattaneo E, Hackam A, Sharp A, Thornberry N, Nicholson DW, Bredesen DE, Hayden MR (2000) Inhibiting caspase cleavage of huntingtin reduces toxicity and aggregate formation in neuronal and nonneuronal cells. J Biol Chem 275:19831–19838.

Wu CC, Chawla F, Games D, Rydel RE, Freedman S, Schenk D, Young WG, Morrison JH, Bloom FE (2004) Selective vulnerability of dentate granule cells prior to amyloid deposition in PDAPP mice: digital morphometric analyses. Proc Natl Acad Sci USA 101:7141–146.

Yang F, Sun X, Beech W, Teter B, Wu S, Sigel J, Vinters HV, Frautschy SA, Cole GM (1998) Antibody to caspase-cleaved actin detects apoptosis in differentiated neuroblastoma and plaque-associated neurons and microglia in Alzheimer's disease. Am J Pathol 152:379–389.

Zhao M, Su J, Head E, Cotman CW (2003) Accumulation of caspase cleaved amyloid precursor protein represents an early neurodegenerative event in aging and in Alzheimer's disease. Neurobiol Dis 14:391–403.

Long-Term Potentiation and Aβ: Targeting Aβ Species, Cellular Mechanisms and Putative Receptors

Michael J. Rowan[1], *Igor Klyubin*[1], *William K. Cullen*[1], *NengWei Hu*[1], and *Roger Anwyl*[2]

Summary. The mechanisms of the relatively selective vulnerability of plasticity at excitatory synapses to the disruptive actions of Aβ may provide new insights into novel therapies for Alzheimer's disease (AD). We have examined the ability of exogenously applied and endogenously generated Aβ antibodies to prevent Aβ inhibition of long-term potentiation (LTP) in the rat hippocampus in vivo. Cell-derived oligomers of Aβ and human cerebrospinal fluid containing Aβ oligomers rapidly inhibited LTP at extremely low concentrations. Antibodies that bound Aβ oligomers abrogated the Aβ-induced disruption of LTP by directly neutralizing them in the brain.

Alzheimer's disease pathology is associated with the activation of certain pro-inflammatory oxidative/nitrosative stress-linked cascades and cytokine pathways. We found that the disruption of LTP in vitro by Aβ was prevented by pharmacological inhibition or gene knockout of JNK, p38 MAPK, NADPH oxidase, iNOS, TNFα and type 1 TNF receptors. Recently we also found that agents, including antibodies and a small molecule inhibitor, that bound a putative receptor for Aβ, the αv integrin, could prevent the Aβ-induced inhibition of LTP both in vitro and in vivo.

Our studies provide a window on likely mechanisms of synaptic plasticity disruption in Alzheimer's disease and how they may be successfully targeted using immunological and pharmacological means, potentially long before pathological synaptic or neuronal loss occurs.

Introduction

Early signs of memory disruption in AD include deficits in the free recall of recent personal events and episodes. There is very strong evidence that the hippocampus plays a critical role in the continual but transient storage of such events and in supporting consolidation into long-term memory. This mnemonic function is thought to require a continuous comparison of incoming integrated perceptual content via the entorhinal cortex, with information and related predictive schemata initially stored/generated in the hippocampus. The underlying cellular basis of hippocampus-dependent memory is believed to require activity-dependent persistent changes in synaptic efficacy, i.e., synaptic plasticity (Morris et al. 2003; Whitlock et al. 2006). Similar activity-dependent synaptic plasticity can be induced by conditioning electrical stimulation, the most commonly studied forms being high frequency stimulation (HFS)-induced long-term potentiation (LTP) and low frequency electrical stimulation-induced long-term depression (LTD).

The impairment of hippocampus-dependent memory in early AD is associated with a relatively selective shrinkage of the hippocampus and related structures in the

[1] Department of Pharmacology & Therapeutics, Biotechnology Building, Trinity College Dublin 2, Ireland, email: mrowan@tcd.ie

[2] Department of Physiology, and Trinity College Institute of Neuroscience, Trinity College Dublin 2, Ireland

Selkoe et al.
Synaptic Plasticity and the Mechanism of Alzheimer's Disease
© Springer-Verlag Berlin Heidelberg 2008

medial temporal lobe. Although the cause of such early shrinkage is by no means certain, the reduction in size and associated enlargement of periventricular space are predictive of clinical progression of the disease. The reason for the early vulnerability of the hippocampus in AD is not well understood. It seems unlikely to be due solely to a disease-specific mechanism and probably reflects the inherent vulnerability of the hippocampus to different insults. This intrinsic vulnerability is attributable to the nature of hippocampal circuitry, which includes a high density of glutamatergic synapses that appear to be continuously undergoing activity-dependent change. The loss of hippocampal synapses in AD may thus be an exaggeration of physiological use-dependent synaptic depression and related synaptic pruning, as has been observed following certain LTD induction protocols (Shinoda et al. 2005).

Because Aβ forms extensive deposits in the hippocampus of AD patients, there is great interest in the possibility that Aβ may be the cause of hippocampal dysfunction and pathology. Aβ can be released in an activity-dependent manner from hippocampus and may function as a negative feedback signal in the control of synaptic transmission, promoting LTD-like mechanisms (Cirrito et al. 2005; Hsieh et al. 2006; Kamenetz et al. 2003). Although LTD may be essential for certain forms of memory formation and under normal circumstances should counter-balance LTP, allowing optimal conditions for information storage in the hippocampal network, an imbalance between these forms of plasticity will have disastrous consequences for memory mechanisms and ultimately for macro-structure.

This review focuses on evidence from our research on the species of Aβ that inhibits LTP in the hippocampus and the mechanisms of this inhibition at the molecular and cellular levels.

LTP in the Hippocampus in Vitro and in Vivo

There is now an extensive and ever-growing knowledge base concerning the multiple mechanisms available for induction and expression of LTP at the different glutamatergic synapses in the hippocampus (e.g., Peineau et al. 2007; Tzingounis and Nicoll 2006; Ward et al. 2006). LTP of AMPA receptor-mediated synaptic transmission at the most extensively studied synapse, between the terminals of Schaffer collaterals from CA3 neurons onto the dendritic spines of CA1 pyramidal neurons, can be induced with several different conditioning protocols, the most common being HFS. Brief periods of HFS transiently open Ca^{2+} permeable NMDA receptors, causing a localized postsynaptic rise in intracellular Ca^{2+} that activates a range of kinases, most notably αCaMKII. Such kinase activity causes a cascade of events to induce synapse-specific increases in AMPA receptor expression and rapid-onset synaptic potentiation. The persistence of the synaptic potentiation depends on many factors, including local dendritic and central somatic protein synthesis. The induction of certain forms of LTP causes extensive synaptic remodeling with concerted pre- and postsynaptic structural changes (Popov et al. 2004).

Using both in vivo and in vitro approaches in our experiments to study the effects of Aβ on LTP has advantages over the use of either alone. The former approach is especially helpful in aiding the extrapolation of our findings to the clinical situation, whereas the latter approach is critical for the elucidation of detailed mechanisms. In the in vivo studies, we implant wire electrodes in the stratum radiatum of the dorsal hippocampus of anesthetized rats and Aβ is injected acutely via a cannula in the ipsilateral lateral cerebral ventricle. In our in vitro studies, submerged hippocampal slices from either mice or rats are superfused acutely with Aβ in a physiological salt solution containing the $GABA_A$ receptor/channel antagonist picrotoxin and electrodes are usually placed in the medial perforant pathway of the dentate gyrus. Electrical stimulation of afferent fibers evokes glutamate release and AMPA receptor-mediated postsynaptic depolarization is recorded as a field excitatory postsynaptic potential (EPSP). The amplitude or slope of the EPSP is used as the measure of synaptic efficacy. To study NMDA receptor-dependent LTP, we apply brief trains of conditioning stimulation at 200 Hz.

LTP-Disrupting Aβ Species: Aβ Oligomers

Some years ago we reported that synthetic Aβ, especially Aβ42, inhibited LTP in vivo at a dose and time that did not affect basal excitatory synaptic transmission (Cullen et al. 1997). Subsequently we have characterized the Aβ species that is primarily responsible for this rapid and potent selective disruption of synaptic plasticity. Using synthetic Aβ containing both soluble and fibrillar species, we found that the soluble species maintained the high potency. Furthermore, monomeric synthetic Aβ was apparently inactive. This finding suggests that a soluble, presumably oligomeric form of Aβ is particularly powerful at inhibiting LTP induction, consistent with the work of Klein et al. on ADDLs (Aβ-derived diffusible ligands; Wang et al. 2002). As synthetic Aβ exists in a wide range of difficult-to-control, and possibly unnatural, assembly states, it was important to also assess the effect of naturally secreted Aβ. We used conditioned medium from Chinese Hampster Ovary cells transfected with amyloid precursor protein with a familial AD mutation (7PA2 cells) that contains Aβ monomers and low-n oligomers (mainly dimers and trimers) but no detectible larger Aβ species. This medium inhibited LTP in a manner similar to soluble synthetic Aβ (Walsh et al. 2002). Moreover, an Aβ oligomer-enriched solution, produced by fractionation of the conditioned medium using size exclusion chromatography, strongly inhibited LTP whereas a monomer-enriched solution had no observable effect (Klyubin et al. 2005).

When tested in vitro, the potency of the naturally secreted Aβ was found to be at least 200-fold greater than synthetic Aβ42, with a threshold concentration of less than ~1 nM and ~200 nM, respectively (Wang et al. 2004b), the latter being similar to what has been reported for synthetic ADDLs (Wang et al. 2002). Indeed Townsend et al. (2006) have reported that naturally secreted Aβ trimers can inhibit LTP in vitro at a concentration of ~100 pM. The reason for the relatively high potency of naturally secreted Aβ is likely to be the stabilization of a particularly active oligomeric species in the conditioned medium that may be in lower abundance in synthetic solutions.

Aβ in Human Cerebrospinal Fluid (CSF) Inhibits LTP

Very recently we have examined the ability of human CSF containing Aβ oligomers to inhibit LTP brain (Klyubin et al., unpublished observations). Several samples of human CSF from cognitively normal individuals and patients with AD were found to contain low-n oligomers (mainly dimers). Human CSF samples with clearly detectible Aβ oligomer content inhibited the induction of LTP whereas the same samples in which Aβ had been previously removed by immunoprecipitation failed to affect LTP induction.

Targeting the Synaptic Plasticity Disrupting Aβ Oligomers: Potential Immunotherapy

We investigated if it was possible to directly neutralize Aβ oligomers in the brain using endogenously generated and exogenously applied antibodies (Klyubin et al. 2005). Anti-Aβ monoclonal antibodies, either the N-terminus directed 6E10 or the mid-region directed 4G8, that avidly bind Aβ oligomers in the conditioned medium from 7PA2 cells, were chosen. LTP was no longer inhibited when the antibody was injected via the cannula at the same time or after the cell-secreted Aβ. Boiled antibody or an isotype control antibody were without effect on the inhibition of LTP. These findings help explain the apparent paradox that antibodies to Aβ can rapidly reverse the memory deficits in transgenic mouse models of AD without reducing brain Aβ burden (Dodart et al. 2002; Kotilinek et al. 2002). It is likely that, in addition to direct scavenging in the CSF, antibodies reach the interstitial fluid in brain regions such as the hippocampus (Bard et al. 2000; Chauhan et al. 2001), where Aβ oligomers can be neutralized, thereby preventing the disruption of synaptic plasticity.

We also evaluated if vaccination of rats with a mixture of pre-aggregated Aβ40 and Aβ42 could prevent the inhibition of LTP induction by naturally produced soluble Aβ (Klyubin et al. 2005). Vaccinated animals that were successfully immunized, as indicated by the presence of plasma antibodies to Aβ, were partially protected from the inhibition of LTP. Importantly, those animals with plasma antibodies with a moderate-to-strong ability to immunoprecipitate Aβ dimers and trimers from 7PA2 conditioned medium were the ones partially resistant to the block of LTP. In contrast, animals with relatively high titers that mainly bound plaque Aβ or recognized Aβ monomers but not oligomers were not protected. Thus endogenously generated antibodies with the ability to bind Aβ low-n oligomers can also neutralize the synaptic plasticity-disrupting effect of natural Aβ.

Very recently we tested the feasibility of using systemic passive immunization to neutralize Aβ oligomers in the brain (Klyubin et al., unpublished observations). Intracardiac pre-injection of the antibody 6E10 prevented the inhibition of LTP caused by subsequent i.c.v. injection of an Aβ oligomer-enriched fraction of 7PA2 conditioned medium. In contrast, systemic injections of a control antibody failed to affect the inhibition of LTP. Furthermore, intracardiac pre-injection of another anti-Aβ antibody, 4G8, prevented the disruption of synaptic plasticity by human CSF containing Aβ oligomers, strongly supporting the potential therapeutic relevance of this approach.

Cellular Stress and Pro-Inflammatory Mechanisms

Cellular stress involving the production of free radicals and inflammatory mediators is strongly implicated in AD (Akiyama et al. 2000; Barnham et al. 2004). Increased activity of certain key stress-activated protein kinases and mitogen-activated protein (MAP) kinases, including p38 MAP kinase (p38 MAPK; Zhu et al. 2004) and c-Jun N-terminal kinase (JNK; Hensley et al., 1999), is known to occur in AD. The ability of inhibitors of these kinases to affect the blocking of LTP by Aβ was assessed in vitro. Aβ failed to inhibit LTP induction after pretreatment with selective inhibitors of p38 MAPK (SB203580) or JNK (JNKI and SP600125) at concentrations that did not affect the induction of LTP when applied alone. In contrast, an inhibitor of p42/44 MAPK, U0126, did not prevent the inhibition of LTP by Aβ, consistent with a selective role for certain stress-activated kinases.

Although an involvement of stress-activated MAP kinases is strongly indicative of glial involvement, these findings are also consistent with a possible role for an activity-dependent mGluR-mediated activation of p38 MAP kinase and JNK, which might directly disrupt LTP. Indeed the group I–II mGluR antagonist LY341495 and the selective mGluR5 antagonist methyl-6-(phenylethynyl)pyridine (MPEP) prevented the Aβ inhibition of LTP in vitro, thus demonstrating the involvement of mGluR5 in the blocking of LTP by Aβ. It is possible that mGluRs located both on glia and neurons may mediate the p38 MAP kinase- and JNK-dependence of the Aβ-mediated blocking of LTP. Co-activation of mGluRs may also lower the threshold for stress kinase activation by Aβ (Kim et al. 2004).

Reactive Oxygen and Nitrogen Free Radicals

The deleterious effects of cellular stress following activation of p38 MAP kinase and JNK are often mediated through free radicals, including nitric oxide (NO) and superoxide. We investigated the role of NO in mediating the blocking of LTP by Aβ in vitro using inhibitors of the enzyme-inducible NO synthase (iNOS) and iNOS knockout mice. The selective iNOS inhibitors 1400W and aminoguanidine significantly reduced the inhibition of LTP by Aβ. Moreover, in iNOS knockout mice, LTP equal to control magnitude was induced in the presence of Aβ.

The neurotoxicity of NO is often due to reaction with superoxide to form the highly reactive peroxynitrite free radical. We examined the ability of the superoxide scavenger superoxide dismutase (SOD) in combination with the hydrogen peroxide scavenger catalase to affect the inhibition of LTP by Aβ. This combination of scavengers prevented the Aβ inhibition of LTP induction, thus supporting a role for superoxide. Since NADPH oxidase is a common source of superoxide, we also tested an inhibitor of this enzyme, diphenyleneiodinium (DPI), and found that it partially abrogated the blocking of LTP by Aβ.

Stress kinase activation and the production of reactive oxygen and nitrogen free radicals are major functions of microglia. A likely involvement of microglia in the inhibition of LTP by Aβ was evaluated by pretreating hippocampal slices with minocycline, an agent that has been found to rapidly and relatively selectively inhibit microglial activation (Tikka et al. 2001; Yrjanheikki et al. 1998; Zhu et al. 2002). Consistent with a requirement for glia-dependent mechanisms, the inhibitory effect of Aβ on LTP induction was prevented by minocycline.

A Critical Role for TNFα in Mediating the Inhibition of LTP by Aβ

Since Aβ is known to induce release of TNFα from microglia (Chao et al. 1994; Meda et al. 1995), we tested the idea that the Aβ-mediated inhibition of LTP occurs as a consequence of release of endogenous TNFα using agents that bind and neutralize TNFα and that inhibit the production of TNFα. Since many toxic effects of TNFα are caused by activation of type 1 TNF receptors (TNF-R1), we also determined whether Aβ-mediated inhibition of LTP induction occurred in mutant mice null for TNF-R1.

Pretreatment with two TNFα ligands, infliximab (a chimeric IgG1κ monoclonal antibody containing a murine TNF-alpha binding region and a human IgG1 backbone) and a TNFα peptide antagonist (an analog of the amino acid sequence 159-178 of the human TNF-R1 with specific and high affinity binding to TNFα), at doses that did not alter LTP induction when applied alone, prevented Aβ inhibition of LTP induction. Thalidomide, which inhibits TNFα production (Majumdar et al. 2002; Sampaio et al. 1991), also selectively prevented the disruptive effect of Aβ.

The pro-inflammatory effects of TNFα are known to be mediated via the TNF receptor, TNF-R1. In mutant mice null for TNF-R1 (TNF-R1-/-), Aβ no longer significantly inhibited the induction of LTP. Moreover, exogenously applied TNFα, like Aβ, had an inhibitory effect on LTP induction that was abrogated in TNF-R1-/- mice and that was mediated through an mGluR5 and p38 MAPK-dependent mechanism.

Recently we examined the potential of targeting TNFα in vivo (Hu et al., unpublished observations). Similar to the in vitro findings described above, i.c.v. injection of either infliximab or the TNFα antagonist peptide prevented the inhibition of LTP caused by subsequent injection of Aβ. As thalidomide is a drug that can be used in humans under controlled conditions, we examined the effectiveness of systemic treatment against the synaptic plasticity-disrupting effect of Aβ. The injection i.c.v. of Aβ failed to inhibit LTP in animals that received peripheral administration of thalidomide.

Putative Aβ Receptors

The Aβ oligomer conformation that rapidly inhibits LTP induction is presumably a pathological assembly state. Its high potency points to an interaction of an epitope in oligomeric Aβ with a specific binding domain(s) of a receptor(s) directly or indirectly regulating synaptic plasticity. Many different receptors for Aβ have been reported, but it is unclear which, if any of them, bind the Aβ oligomer species that causes the inhibition of LTP at extremely low concentrations.

The Aβ-induced inhibition of LTP could involve direct Aβ binding and activation of TNFR1, as Aβ was found to bind to TNFR1 with high affinity (Li et al. 2004). Importantly, the species of Aβ that bound to TNFR1 was found to be the soluble oligomers rather than insoluble fibrils (Li et al. 2004). Thus it is possible that TNF-R1 is activated both by endogenously released TNF and by Aβ.

An Aβ-induced release of endogenous TNF from microglia is likely to be mediated by the binding of Aβ to a receptor located on microglia, which apart from TNFR1 potentially include a scavenger receptor (El Khoury et al. 1996), a receptor for advanced

Table 1. Summary of the effects of different treatments on the inhibition of LTP by Aβ both in vivo and in vitro (see text for references)

System	Effect on inhibition of LTP	
	Abrogation	No effect
Synthetic Aβ in vitro (monomers, oligomers, protofibrils and fibrils)	JNK inhibitor Cdk5 inhibitor p38 MAPK inhibitor Minocycline α7 nicotinic ACh receptor antagonist mGluR 5 antagonist iNOS inhibitor iNOS knock-out Superoxide scavenger NADPH oxidase inhibitor TNF-R1 knock-out TNFα scavengers TNFα production inhibitor Actin stabilizer Anti-αv integrin antibodies αv integrin antagonist Superfibronectin Echistatin	p42-p44 MAPK inhibitor
Synthetic soluble Aβ in vivo (fibril-free)	Anti-αv integrin antibody αv integrin antagonist TNFα scavengers	
7PA2 cell medium *in vivo* (Aβ *low n-oligomers* ± monomers)	Active Aβ immunization (partially effective) Passive immunization with anti-Aβ antibody (pre-Aβ administration, co-administration and post-Aβ administration)	
7PA2 cells in vitro (Aβ low n-oligomers and monomers) prior to *in vivo* administration	γ-secretase inhibitor	Insulin-degrading enzyme
Human CSF in vivo (monomers and dimers)	Passive immunization with anti-Aβ antibody	

glycation end product (Yan et al. 1996) or a receptor complex including an integrin protein (Bamberger et al. 2003).

Recent evidence supports a critical role for integrins as potential neuronal and glia receptors for Aβ-induced cytotoxicity (Caltagarone et al. 2007). Using selective integrin-blocking antibodies, Wright et al. (2007) showed that both the binding/deposition and neurotoxicity of Aβ in human cortical primary neurons are mediated through αvβ1and α2β1 integrins. Consequently, we investigated if selective αv integrin-blocking antibodies and a small molecule inhibitor of αv integrins can prevent the Aβ inhibition of LTP both in vitro and in vivo (Wang et al. 2007). We also examined the ability of other integrin ligands including echistatin and superfibronectin to block the Aβ inhibition of LTP in vitro.

Perfusion of three different antibodies to αv-containing integrins, 18C7, 20A9 and 17E6, none of which affected LTP induction alone, prevented the Aβ-mediated inhibition of LTP in hippocampal slices. In contrast, two isotype control antibodies did not significantly affect the Aβ inhibition of LTP. Similarly, in vivo, pre-treatment of anesthetized rats with the αv integrin antibody 17E6 but not an isotype control antibody prevented the inhibition of LTP caused by i.c.v. injection of soluble fibril-free Aβ.

Consistent with a requirement for integrins, both superfibronectin, which is a ligand for αvβ1 (Morla et al. 1994), and echistatin, which is a snake venom-derived disintegrin that can inhibit RGD-dependent integrin, prevented the Aβ-mediated block of LTP at a concentration that alone did not significantly affect LTP induction.

Next we tested SM256, a non-peptide small molecule that is a potent αv antagonist and that inhibits αv-mediated cell adhesion (Van Waes et al. 2000). SM256 applied in vitro in the perfusion medium prevented the Aβ-mediated block of LTP without affecting control LTP. Importantly, in in vivo experiments, systemic (i.p.) pre-administration of SM256 abrogated the inhibition of LTP caused by i.c.v. injection of soluble, fibril-free Aβ.

Conclusion

Table 1 summarizes the effects of different treatments on the inhibiton of LTP by Aβ both in vitro and in vivo. One possible scenario is that Aβ oligomers bind αv integrin to trigger TNFα release, which in turn causes activation of stress kinases and down-stream production of nitric oxide, superoxide and other mediators that compromise normal, activity-dependent physiological activation of these cellular pathways by synaptically released glutamate acting through NMDA receptors and mGlu 5 receptors, resulting in inhibition of high frequency stimulation-induced LTP and perhaps promotion of LTD. Many of the key factors mediating the rapid inhibition of LTP by Aβ, including the involvement of oligomers and cellular stress and LTD-like mechanisms, have been implicated in Aβ-induced synaptic pruning and neurodegeneration. Our studies provide insights into how synaptic plasticity disruption may be successfully targeted using immunological and pharmacological means, potentially long before pathological synaptic or neuronal loss occurs in AD.

Acknowledgements. Supported by Science Foundation Ireland and Programme for Research in Third Level Institutions in Ireland. Thanks to Prof. Dennis Selkoe and Dominic Walsh for extensive collaboration.

References

Akiyama H, Barger S, Barnum S, Bradt B, Bauer J, Cole GM, Cooper NR, Eikelenboom P, Emmerling M, Fiebich BL, Finch CE, Frautschy S, Griffin WS, Hampel H, Hull M, Landreth G, Lue L, Mrak R, Mackenzie IR, McGeer PL, O'Banion MK, Pachter J, Pasinetti G, Plata-Salaman C, Rogers J, Rydel R, Shen Y, Streit W, Strohmeyer R, Tooyoma I, Van Muiswinkel FL, Veerhuis R, Walker D, Webster S, Wegrzyniak B, Wenk G, Wyss-Coray T (2000) Inflammation and Alzheimer's disease. Neurobiol Aging 21:383–421

Bamberger ME, Harris ME, McDonald DR, Husemann J, Landreth GE (2003) A cell surface receptor complex for fibrillar-amyloid mediates microglial activation. J Neurosci 23:2665–2674

Bard F, Cannon C, Barbour R, Burke RL, Games D, Grajeda H, Guido T, Hu K, Huang J, Johnson-Wood K, Khan K, Kholodenko D, Lee M, Lieberburg I, Motter R, Nguyen M, Soriano F, Vasquez N, Weiss K, Welch B, Seubert P, Schenk D, Yednock T (2000) Peripherally administered antibodies against amyloid beta-peptide enter the central nervous system and reduce pathology in a mouse model of Alzheimer disease. Nature Med 6:916–919

Barnham KJ, Masters CL, Bush AI (2004) Neurodegenerative diseases and oxidative stress. Nature Rev Drug Discov 3:205–214

Caltagarone J, Jing Z, Bowser R (2007) Focal adhesions regulate Abeta signaling and cell death in Alzheimer's disease. Biochim Biophys Acta 1772:438–445

Chao CC, Hu S, Kravitz FH, Tsang M, Anderson WR, Peterson PK (1994) Transforming growth factor-beta protects human neurons against beta-amyloid-induced injury. Mol Chem Neuropathol 23:159–78

Chauhan NB, Siegel GJ, Lichtor T (2001) Distribution of intraventricularly administered antiamyloid-beta peptide (Abeta) antibody in the mouse brain. J Neurosci Res 66:231–235

Cirrito JR, Yamada KA, Finn MB, Sloviter RS, Bales KR, May PC, Schoepp DD, Paul SM, Mennerick S, Holtzman DM (2005) Synaptic activity regulates interstitial fluid amyloid-beta levels *in vivo*. Neuron 48:913–922

Cullen WK, Suh Y-H, Anwyl R, Rowan MJ (1997) Block of LTP in rat hippocampus *in vivo* by beta-amyloid precursor protein fragments. NeuroReport 8:3213–3217

Dodart JC, Bales KR, Gannon KS, Greene SJ, DeMattos RB, Mathis C, DeLong CA, Wu S, Wu X, Holtzman DM, Paul SM (2002) Immunization reverses memory deficits without reducing brain Abeta burden in Alzheimer's disease model. Nature Neurosci 5:452–457

El Khoury J, Hickman SE, Thomas CA, Cao L, Silverstein SC, Loike JD (1996) Scavenger receptor-mediated adhesion of microglia to beta amyloid fibrils. Nature 382:716–719

Hensley K, Floyd RA, Zheng NY, Nael R, Robinson KA, Nguyen X, Pye QN, Stewart CA, Geddes J, Markesbery WR, Patel E, Johnson GV, Bing G (1999) p38 kinase is activated in the Alzheimer's disease brain. J Neurochem 72:2053–2058

Hsieh H, Boehm J, Sato C, Iwatsubo T, Tomita T, Sisodia S, Malinow R (2006) AMPAR removal underlies Abeta-induced synaptic depression and dendritic spine loss. Neuron 52:831–843

Kamenetz F, Tomita T, Hsieh H, Seabrook G, Borchelt D, Iwatsubo T, Sisodia S, Malinow R (2003) APP processing and synaptic function. Neuron 37:925–37

Kim SH, Smith CJ, Van Eldik LJ (2004) Importance of MAPK pathways for microglial pro-inflammatory cytokine IL-1β production. Neurobiol Aging 25:431–439

Klyubin I, Walsh DM, Lemere CA, Cullen WK, Shankar GM, Betts V, Spooner ET, Jiang L, Anwyl R, Selkoe DJ, Rowan MJ (2005) Amyloid beta protein immunotherapy neutralizes Abeta oligomers that disrupt synaptic plasticity *in vivo*. Nature Med 11:556–561

Kotilinek LA, Bacskai B, Westerman M, Kawarabayashi T, Younkin L, Hyman BT, Younkin S, Ashe KH (2002) Reversible memory loss in a mouse transgenic model of Alzheimer's disease. J Neurosci 22:6331–6335

Li R, Yang L, Lindholm K, Konishi Y, Yue X, Hampel H, Zhang D, Shen Y (2004) Tumor necrosis factor death receptor signaling cascade is required for amyloid-beta protein-induced neuron death. J Neurosci 24:1760–1771

Majumdar S, Lamothe B, Aggarwal BR (2002) Thalidomide suppresses NFkB activation induced by TNF and H2O2, but not that activated by ceramide, lipopolysaccharides or phorbol ester. J Immunol 168:2644–2651

Meda L, Cassatella MA, Szendrei GI, Otvos L Jr, Baron P, Villalba M, Ferrari D, Rossi F (1995) Activation of microglial cells by beta-amyloid protein and interferon-gamma. Nature 374:647–50

Morla A, Zhang Z, Ruoslahti E (1994) Superfibronectin is a functionally distinct form of fibronectin. Nature 367:193–196

Morris RGM, Moser EI, Riedel G, Martin SJ, Sandin J, Day M, O'Carroll C (2003) Elements of a neurobiological theory of the hippocampus: the role of activity-dependent synaptic plasticity in memory. Philos Trans R Soc Lond B Biol Sci 358:773–786

Peineau S, Taghibiglou C, Bradley C, Wong TP, Liu L, Lu J, Lo E, Wu D, Saule E, Bouschet T, Matthews P, Isaac JT, Bortolotto ZA, Wang YT, Collingridge GL (2007) LTP inhibits LTD in the hippocampus via regulation of GSK3beta. Neuron 53:703–717

Popov VI, Davies HA, Rogachevsky VV, Patrushev IV, Errington ML, Gabbott PL, Bliss TV, Stewart MG (2004) Remodelling of synaptic morphology but unchanged synaptic density during late phase long-term potentiation (LTP): a serial section electron micrograph study in the dentate gyrus in the anaesthetised rat. Neuroscience 128:251–262

Sampaio EP, Sarno EN, Galilly R, Cohn ZA, Kaplan G (1991) Thalidomide selectively inhibits tumor necrosis factor alpha production by stimulated human monocytes. J Exp Med 173:699–703

Shankar GM, Bloodgood BL, Townsend M, Walsh DM, Selkoe DJ, Sabatini BL (2007) Natural oligomers of the Alzheimer amyloid-beta protein induce reversible synapse loss by modulating an NMDA-type glutamate receptor-dependent signaling pathway. J Neurosci 27:2866–2875

Shinoda Y, Kamikubo Y, Egashira Y, Tominaga-Yoshino K, Ogura A (2005) Repetition of mGluR-dependent LTD causes slowly developing persistent reduction in synaptic strength accompanied by synapse elimination. Brain Res 1042:99–107

Tikka T, Fiebich B, Goldsteins G, Keinanen R, Koistinaho J (2001) Minocycline, a tetracycline derivative, is neuroprotective against excitiotoxicity by inhibiting activating and proliferation of microglia. J Neurosc 21:2580–2588

Townsend M, Shankar GM, Mehta T, Walsh DM, Selkoe DJ (2006) Effects of secreted oligomers of amyloid beta-protein on hippocampal synaptic plasticity: a potent role for trimers. J Physiol 572:477–492

Tzingounis AV, Nicoll RA (2006) Arc/Arg3.1: linking gene expression to synaptic plasticity and memory. Neuron 52:403–407

Van Waes C, Enamorado-Ayala I, Hecht D, Sulica L, Chen Z, Batt DG, Mousa S (2000) Effects of the novel alphav integrin antagonist SM256 and cis-platinum on growth of murine squamous cell carcinoma PAM LY8. Int J Oncol 16:1189–1195

Walsh DM, Klyubin I, Fadeeva JV, Cullen WK, Anwyl R, Wolfe MS, Rowan MJ, Selkoe DJ (2002) Naturally secreted oligomers of amyloid β protein potently inhibit hippocampal long-term potentiation *in vivo*. Nature 416:535–539.

Wang HW, Pasternak JF, Kuo H, Ristic H, Lambert MP, Chromy B, Viola KL, Klein WL, Stine WB, Krafft GA, Trommer BL (2002) Soluble oligomers of beta amyloid (1-42) inhibit long-term potentiation but not long-term depression in rat dentate gyrus. Brain Res 924:133–40

Wang Q, Rowan MJ, Anwyl R (2004a) Beta-amyloid-mediated inhibition of NMDA receptor-dependent long-term potentiation induction involves activation of microglia and stimulation of inducible nitric oxide synthase and superoxide. J Neurosci 24:6049–6056

Wang Q, Walsh DM, Rowan MJ, Selkoe DJ, Anwyl R (2004b) Block of long-term potentiation by naturally secreted and synthetic amyloid beta-peptide in hippocampal slices is mediated via activation of the kinases c-Jun N-terminal kinase, cyclin-dependent kinase 5, and p38 mitogen-activated protein kinase as well as metabotropic glutamate receptor type 5. J Neurosci 24:3370–3378

Wang Q, Klyubin I, Wright S, Griswold-Prenner I, Rowan MJ, Anwyl R (2007) AlphaV integrins mediate beta-amyloid induced inhibition of long-term potentiation. Neurobiol Aging, in press

Ward B, McGuinness L, Akerman CJ, Fine A, Bliss TV, Emptage NJ (2006) State-dependent mechanisms of LTP expression revealed by optical quantal analysis. Neuron 52:649–61

Whitlock JR, Heynen AJ, Shuler MG, Bear MF (2006) Learning induces long-term potentiation in the hippocampus. Science 313:1093–1097

Wright S, Malinin NL, Powell KA, Yednock T, Rydel RE, Griswold-Prenner I (2007) Alpha2beta1 and alphaVbeta1 integrin signaling pathways mediate amyloid-beta-induced neurotoxicity. Neurobiol Aging 28:226–37

Yan SD, Chen X, Fu J, Zhu H, Roher A, Slattery T, Morser J, Nawroth P, Stern D, Schmidt AM (1996) RAGE and amyloid-peptide neurotoxicity in Alzheimer's disease. J Biol Chem 274:15493–15499

Yrjanheikki J, Keinnanen R, Pellikka M, Hokfelt T, Koistinaho J (1998) Tetracyclines inhibit microglial activation and are neuroprotective in global brain ischaemia. Proc Natl Acad Sci USA 95:15769–15774

Zhu S, Stavrovskaya IG, Drozda M, Kim BY, Ona V, Li M, Sarang S, Liu AS, Hartley DM, Wu du C, Gullans S, Ferrante RJ, Przedborski S, Kristal BS, Friedlander RM (2002) Minocycline inhibits cytochrome c release and delays progression of amyotrophic lateral sclerosis in mice. Nature 417:74–78

Zhu X, Raina AK, Lee HG, Casadesus G, Smith MA, Perry G (2004) Oxidative stress signalling in Alzheimer's disease. Brain Res 1000:32–39

Genes, Synapses and Autism Spectrum Disorders

Thomas Bourgeron[1]

Summary. Autism spectrum disorders (ASD) can be the consequences of genetic anomalies affecting distinct physiological processes such as chromatin remodeling (MECP2), synaptic gene regulation (FMRP), actin skeleton dynamics (TSC1/TSC2, NF1), cell growth (PTEN) and calcium signaling (CACNA1C). Although the causative genes are numerous and diverse, they might all interfere with a more restricted number of downstream pathways at the origin of ASD. One such pathway includes the synaptic cell adhesion molecules, NLGN3, NLGN4X, and NRXN1, and a postsynaptic scaffolding protein, SHANK3. This protein complex was shown to be crucial for maintaining functional synapses as well as the adequate balance between neuronal excitation and inhibition. In this review, we report genetic and neurobiological findings that highlight the major role of synaptic genes in the susceptibility to ASD. Based on these pieces of evidence, we propose that future studies should include the modulation of synaptic function as a focus for functional analyses and the development of new therapeutic strategies.

Introduction

Thanks to a highly complex phylogenetic and ontogenic process, humans have the ability to communicate with language and highly specialized skills to recognize social cues (eye gaze, joint attention, theory of mind, empathy). In some individuals, this ability to communicate is hampered by the occurrence of genetic/epigenetic variations and/or environmental insults. After the exclusion of known biological diseases (e.g., deafness) and known environmental causes (e.g., social, teaching), approximately 8 to 10% of school age children suffer from language and/or communication difficulties (Shaywitz et al. 1990; Tomblin et al. 1997; Baird et al. 2006). One of the most severe syndromes associated with an alteration of language and social communication is autism.

Autism was first described by the psychiatrist Leo Kanner in 1943 and is diagnosed on the basis of three behaviorally altered domains, namely social deficits, impaired language and communication, and stereotyped and repetitive behaviors (Kanner 1943). Beyond this unifying definition lies an extreme degree of clinical heterogeneity, ranging from debilitating impairments to mild personality traits. Hence autism is surely not a single disease entity but rather a complex phenotype encompassing either multiple "autistic disorders" or a continuum of autistic-like traits and behaviors. To take into consideration this heterogeneity, the term autism spectrum disorders (ASD) is now used and includes autistic syndrome, pervasive development disorder not otherwise specified (PDD-NOS), Asperger syndrome, childhood disintegrative disorder (CDD),

[1] Institut Pasteur, Université Denis Diderot Paris 7, 25 rue du Docteur Roux, 75724 Paris Cedex 15, France, email: thomasb@pasteur.fr

Selkoe et al.
Synaptic Plasticity and the Mechanism of Alzheimer's Disease
© Springer-Verlag Berlin Heidelberg 2008

and Rett syndrome (APA 1994). While CDD and Rett syndrome are severe neurological disorders, Asperger syndrome refers to the portion of the ASD continuum characterized by higher cognitive abilities and by more normal language function.

The behavioral singularities that occur in ASD are related to a wide spectrum of cognitive functions such as language, memory, and visual and auditive attentions. Two of these cognitive deficits are rather characteristic of ASD: a weak "central coherence" and the lack of a "theory of mind". "Central coherence" defines our ability to understand context. Individuals with ASD are sometimes better than age-matched controls at detecting details in a picture, but they have great difficulty in seeing "the bigger picture" and in understanding the context of the situation (Frith 1998). The term "theory of mind" was coined by Premak and Woodruff (1978) to describe an individual's understanding of the motives, knowledge, and beliefs of others. Individuals with ASD have an absence of theory of mind or a delay in the acquisition of it. Hence, this deficit could be a major cause of their difficulties in social interactions (Baron-Cohen et al. 1985).

Epidemiologic studies have reported a dramatic rise of ASD during the last two decades (from 2–5 to 60 in 10 000 children), but this recent increase is most likely explained by the use of broader diagnostic criteria and the increased attention of the medical community (Fombonne 2005). For still unknown reasons, males are more frequently affected than females. The male to female ratio is 4:1, but increases to 23:1 in individuals without identified morphological or brain abnormalities (Miles and Hillman 2000).

Genetic Causes of Autism Spectrum Disorders

Since the original reports by Kanner and Asperger, many studies have supported a genetic etiology for autism (Freitag 2007). Familial cases are substantially more frequent than expected by chance. Indeed, the recurrence risk of autism in sib-ships is approximately 45 times greater than in the general population. Furthermore, twin studies have documented a higher concordance rate in monozygotic (60–91%) than in dizygotic twins (0–6%; Bailey et al. 1995). However, due to the heterogeneity of the syndrome and the absence of apparent mendelian segregation, the mode of inheritance of ASD is still a matter of debate. The long-time accepted polygenic model was recently challenged by the identification of rare forms of an apparently monogenic form of ASD caused by a single gene mutation or de novo copy number variants (CNVs; Jamain, et al. 2003; Jacquemont, et al. 2006; Durand et al. 2007; Sebat et al. 2007; Szatmari et al. 2007). Although currently restricted to a limited number of patients, these apparent monogenic forms may be a more frequent cause of ASD than originally expected. In addition, epigenetic anomalies are strongly suspected, but their actual impact on autism remains largely unknown.

In approximately 30% of cases, ASD are associated with a known genetic syndrome (e.g., fragile X syndrome, tuberous sclerosis, and neurofibromatosis) or with chromosomal rearrangements (Jacquemont et al. 2006; Vorstman et al. 2006; Freitag 2007; Sebat et al. 2007; Fig. 1). The association with known genetic disorders indicates that anomalies in distinct physiological processes such as chromatin remodeling (Rett syndrome; MECP2), synaptic gene regulation (Fragile X syndrome; FMRP), actin skele-

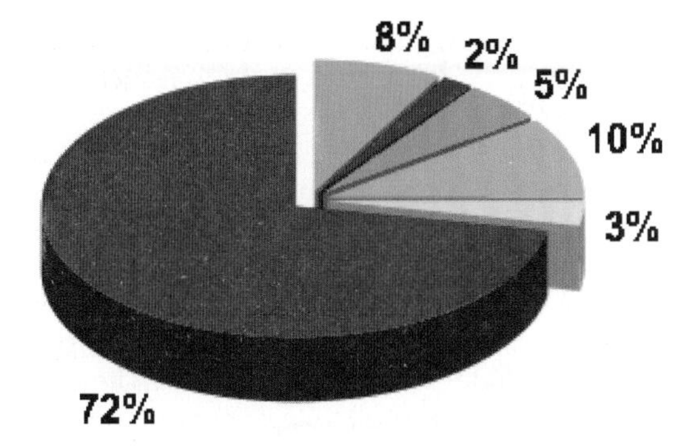

Fig. 1. Broad estimation of the heterogeneous causes of autism spectrum disorders. ASD includes ≈8% of known genetic syndromes (e.g., Fragile X Syndrome (FXS), Tuberous sclerosis (TS), Neurofibromatosis (NF)), ≈2% of Rett syndrome, ≈5% of chromosomal rearrangements (CR), ≈10% of copy number variants (CNVs), ≈3% of mutations in the NLGN/NRXN/SHANK3 pathway, and ≈72% of unknown causes. These numbers are only a broad estimation since epidemiological data concerning the causes of ASD are limited. In addition, the percentages may vary for sporadic or familial cases and if the affected individual has dysmorphic features

ton dynamics (Tuberous sclerosis; TSC1/TSC2, Neurofibromatosis NF1), cell growth (Cowden syndrome; PTEN) and calcium signaling (Timothy syndrome; CACNA1C) can increase the risk for an individual to have ASD. Although the causative genes are numerous and diverse, they might all interfere with a more restricted number of downstream pathways at the origin of ASD. Consistent with this hypothesis, several synaptic genes have been found to be strongly associated with non-syndromic ASD (Fig. 2), providing a better view of the complex pathways that alter properties of the neuronal networks and likely contribute to the disorders (Belmonte and Bourgeron 2006).

Synaptic Genes and Autism Spectrum Disorders

The cell adhesion molecules NLGN and NRXN The first synaptic genes associated with autism and Asperger syndrome were the X-linked neuroligins, *NLGN3* and *NLGN4X*. Neuroligins are cell adhesion molecules with an esterase domain that play a crucial

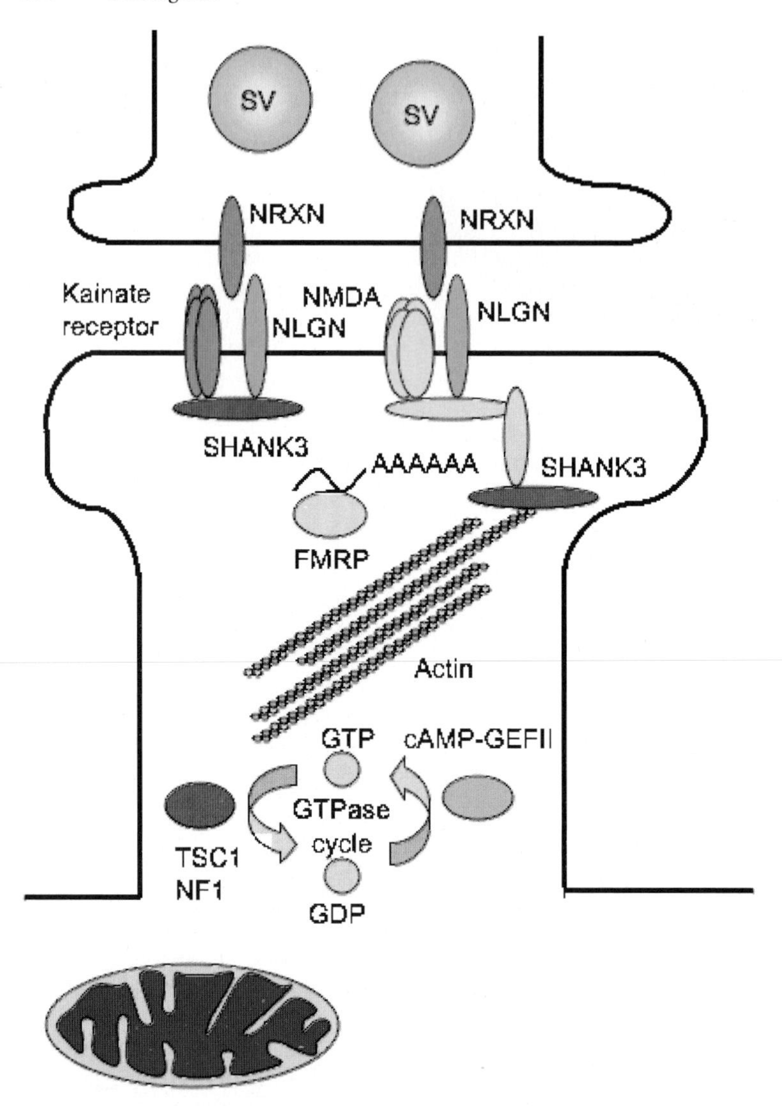

Fig. 2. The synaptic genes associated with ASD. Synaptic vesicles (SV) and neurexins (NRXN) are present at the presynaptic side of a glutamatergic synapse. At the postsynaptic side, the NLGN and the glutamate receptors bind to scaffolding proteins of the postsynaptic density (PSD), such as SHANK3. Additional proteins associated with ASD may also regulate the NRXN/NLGN/SHANK3 pathway. The FMRP controls the translation of several synaptic proteins. TSC1 and NF1 are regulating the actin dynamics and the morphology of the neuron. MECP2 (not shown here) controls gene regulation by modifying chromatin structure

role in the formation of functional synapses. They are located at the postsynaptic side of the synapse and bind to other cell adhesion molecules called neurexins located on the pre-synaptic side of the synapse. There are five neuroligin genes, *NLGN1*, *NLGN2*,

NLGN3, *NLGN4X* and *NLGN4Y*, in the human genome. The first reported mutation of a member of the neuroligin family was a frame-shift mutation in the X-linked *NLGN4X* gene in two brothers, one with autism and the other with Asperger syndrome (Jamain et al. 2003). This mutation (D396X) was located in the esterase domain, leading to premature termination of the protein before the trans-membrane domain. In parallel, a non-synonymous mutation (R451C) of the X-linked *NLGN3*, which affects a highly conserved amino acid from the esterase domain, was identified in a second family with two brothers, one with autism and one with Asperger syndrome. These mutations (NLGN4X D396X and NLGN3 R451C) were found to cause abnormal synaptogenesis in cultured neuronal cells (Chih et al. 2004; Comoletti et al. 2004). Mutations within the *NLGN3* and *NLGN4X* coding sequences are rare (<1% of individuals with ASD; Vincent et al. 2004; Gauthier et al. 2005; Ylisaukko-oja et al. 2005; Blasi et al. 2006). However, other groups have replicated the association by identifying independent *NLGN4X* mutations in individuals with ASD and/or mental retardation (Laumonnier et al. 2004; Yan et al. 2004). In addition, abnormal *NLGN3* and *NLGN4X* spliced isoforms were detected in blood cells from individuals with ASD (Talebizadeh et al. 2006). If these transcripts are actually present in the brain of the affected individuals, this abnormal expression may represent a new type of NLGN alteration.

Using a whole-genome approach, an international collaborative effort – the Autism Genome Project Consortium – investigated 1,168 multiplex families for the presence of linkage and CNVs (Szatmari et al. 2007). This analysis could detect a new locus for autism on chromosome 11 and the presence of a *de novo* deletion of the *NRXN1* gene located on chromosome 2p16 in two sisters with ASD. The *NRXN1* gene encodes neurexin, the presynaptic partner of the NLGN, confirming that a defect in this synaptic pathway could cause ASD.

The synaptic scaffolding protein SHANK3 Among the most frequent chromosomal rearrangements associated with cognitive deficits, the 22q13 microdeletion syndrome is characterized by neonatal hypotonia, global developmental delay, normal to accelerated growth, absent to severely delayed speech, and minor dysmorphic features (Manning et al. 2004). Behavior is described as "autistic-like" and includes high tolerance to pain and habitual chewing or mouthing. Among the three genes (*SHANK3*, *ACR* and *RABL2B*) located in the minimal 130 kb sub-telomeric region, SHANK3 is known to regulate the structural organization of dendritic spines and is a binding partner of NLGN (Boeckers et al. 2002; Meyer et al. 2004).

Durand et al. (2007) sequenced the coding sequence of *SHANK3* in 227 individuals with ASD and showed that mutations, or loss of one copy of the gene, were associated with autism, whereas the presence of an extra copy might be associated with Asperger syndrome. Among the variations identified, a *de novo* frame-shift mutation that originated in a mother with germinal mosaicism was present in two brothers with autism. Expression in cultured hippocampal neurons of the rat *Shank3* cDNA carrying the frame-shift mutation indicated that, unlike the wild-type protein, the truncated protein is absent in the dendritic spines. These results provide further genetic and functional evidence that the synaptic pathway, which includes *NLGN*, *NRXN* and *SHANK3*, is associated with ASD.

Other synaptic gene candidates for susceptibility to autism spectrum disorders The mutations identified in *NLGN*, *NRXN1* and *SHANK3* are segregating with the disorder and are strongly altering the function of these proteins. These features make these genes likely susceptibility genes for ASD. Other synaptic genes are associated with ASD but the functional consequences of the "risk alleles" remain largely unknown. Two of these candidate genes encode glutamate receptor subunits, the *glutamate receptor ionotropic kainate 2* gene (*GRIK2* or *GluR6*) and the N-methyl-D-aspartate (NMDA) receptor 2 subunit (*GRIN2A).* Both proteins are direct (or close) binding partners of the NLGN/NRXN1/SHANK3 protein complex.

GRIK2 is located on chromosome 6q13-21, and a complete loss of function of the gene was recently shown to cause mild to severe learning disabilities (Motazacker et al. 2007). In three independent cohorts of affected individuals, *GRIK2* was associated with ASD. The first sample comprised 59 sib-pairs families and 107 parent-offspring trios from France and Sweden. A significant excess of *GRIK2* allelic sharing ($P = 0.0004$) and maternal transmission disequilibrium ($P = 0.0005$) was observed in affected children (Jamain et al. 2002). The two follow-up studies also found a transmission disequilibrium of *GRIK2* in 174 Chinese Han ($P = 0.01$; Shuang et al. 2004) and 126 Korean ($P < 0.001$) families with ASD (Kim et al. 2007). Interestingly, one amino acid change (M867I), which acts as a gain of function (Strutz-Seebohm et al. 2006), was found in 8% of the autistic subjects and in 4% of the control population and was more maternally transmitted than expected to autistic males ($P = 0.007$; Jamain et al. 2002).

GRIN2A is located on chromosome 16p11-13, a region detected by linkage analyses of families with ASD in several independent studies. Barnby et al. (2005) found a highly significant difference in the distribution of *GRIN2A* haplotypes between cases and controls in both multiplex families ($P = 2 \times 10E{-}11$) and singleton cases ($P = 6 \times 10E{-}7$; Barnby et al. 2005). Interestingly, a "risk" haplotype was seen more frequently in cases than in controls ($P = 4 \times 10E{-}7$), and one "protective haplotype" was seen more frequently in controls than in cases ($P = 0.001$).

Atypical Synapses in Autism Spectrum Disorders?

Taken together, these results strongly suggest that the NRXN/NLGN/SHANK protein complex is a core component of ASD (Fig. 3). Although there are few data about the specific role of this pathway in the human brain, studies on neuronal cell culture and animal models have provided crucial information about its function.

First, neuroligins and neurexins enhance synapse formation in vitro (Scheiffele et al. 2000; Dean et al. 2003) but, surprisingly, are not required for the generation of synapses *in vivo* (Varoqueaux et al. 2006). Indeed, results from knockout (KO) mice demonstrated that neither NLGNs nor neurexins are required for the initial formation of synapses, but both are essential for synaptic function and mouse survival (Missler 2003; Varoqueaux et al. 2006; Dudanova et al. 2007). Therefore, neuroligins may not establish but may specify and validate synapses via an activity-dependent mechanism, with different neuroligins acting on distinct types of synapses (Chubykin et al. 2007). This model proposed by Chubykin et al. (2007), reconciles the overexpression and knockout phenotypes and suggests that neuroligins contribute to the use-dependent formation of neural circuits.

Secondly, neuroligins and neurexins are also emerging as central organizing molecules for excitatory glutamatergic and inhibitory GABAergic synapses in mam-

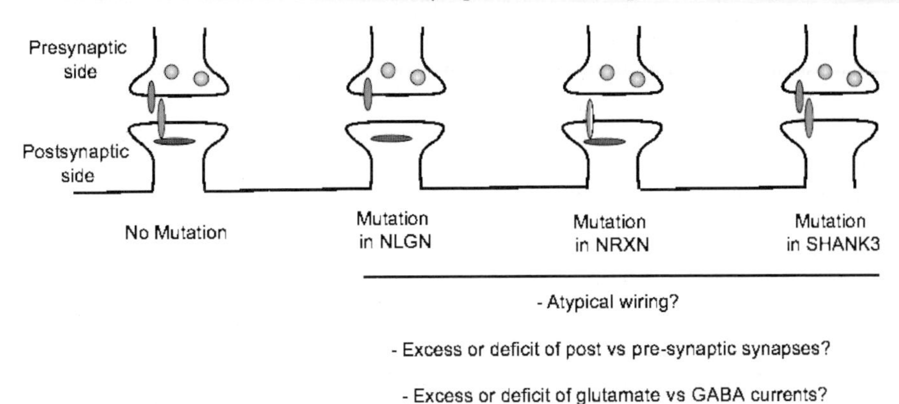

Fig. 3. Models of atypical synapses in ASD. Mutations in NLGN (*green oval*), NRXN (*orange oval*), or SHANK3 (*red oval*) will result in atypical synaptic architecture. Among the possible alterations, these mutations may change the wiring of the brain by modifying the balance between the number of post- vs pre-synapses and/or by causing an imbalance between glutamate and GABA currents

malian brain (Graf et al. 2004; Prange et al. 2004). NLGN1, NLGN3, and NLGN4X are specific to glutamatergic synapses, whereas NLGN2 is restricted to GABAergic synapses (Varoqueaux et al. 2004). This selectivity for glutamatergic vs GABAergic synapses is also conferred by alternative splicing of both neuroligins and neurexins (Chih et al. 2006; Comoletti et al. 2006; Graf et al. 2006; Craig and Kang 2007). Hence, the insertion of an alternative spliced exon in β-neurexins selectively promotes GABAergic synaptic function, whereas the insertion of an alternative spliced exon in neuroligin 1 selectively promotes glutamatergic synaptic function (Chih et al. 2006; Comoletti et al. 2006; Graf et al. 2006). This characteristic of the pathway is highly relevant to ASD since imbalance between excitation and inhibition could lead to epilepsy, a disease observed in almost 25% of individuals with ASD (Tuchman and Rapin 2002).

Finally, at least in humans, this pathway appears to be highly sensitive to gene dosage. Indeed, a deletion of a single copy of *SHANK3* or *NRXN1* seems to be enough to cause ASD (Durand et al. 2007; Szatmari et al. 2007). Therefore, the change in the number and/or the quality of the synaptic contacts may be subtle but the consequences at the cognitive level are severe, since patients sometimes present with a complete absence of speech and great difficulties in social communication. This sensitivity to gene dosage is an important feature that should be taken into account when reconsidering the mode of inheritance of ASD and, more generally, for understanding the evolution of higher brain functions.

Conclusion and Perspectives

Even if genetic and functional data are still sparse and limited, a picture is starting to emerge that, in some cases, ASD may be due to a problem in the development of specific pathways controlling neuronal networks. Many questions remain unanswered, such as the nature of the affected neuronal networks and the potential of reversing the phenotype induced by the mutations. Animalmodels carrying mutations in the NLGN/NRXN/SHANK3 pathway should provide key information. Recent studies on mutant mice for *Fmr1* (fragile X syndrome) and *Mecp2* (Rett syndrome) are encouraging since the neurological phenotype could be reversed in both mice by pharmacological treatment (Hayashi et al. 2007) or reactivation of the gene (Guy et al. 2007). Based on this evidence, the modulation of synaptic function should be one of the priorities for future functional analyses and for the development of new therapeutic strategies for ASD.

Acknowledgements. I would like to thank Bernard Lakowski for helpful discussion and for reading the manuscript. This work was supported by the Pasteur Institute, INSERM, Assistance Publique-Hôpitaux de Paris, Fondation France Télécom, Cure Autism Now, Fondation de France, Fondation Biomédicale de la Mairie de Paris, Fondation pour la Recherche Médicale, EUSynapse European Commission FP6, AUTISM MOLGEN European Commission FP6, ENI-NET European Commission FP6.

References

APA (1994) Diagnostic and statistical manual of mental disorders, Fourth Edition, APA, Washington, DC.

Bailey A, Le Couteur A, Gottesman I, Bolton P, Simonoff E, Yuzda E, Rutter M (1995) Autism as a strongly genetic disorder: evidence from a British twin study. Psychol Med 25:63–77.

Baird G, Simonoff E, Pickles A, Chandler S, Loucas T, Meldrum D, Charman T (2006) Prevalence of disorders of the autism spectrum in a population cohort of children in South Thames: the Special Needs and Autism Project (SNAP). Lancet 368:210–215.

Baron-Cohen S, Leslie AM, Frith U (1985) Does the autistic child have a "theory of mind"? Cognition 21:37–46.

Barnby G, Abbott A, Sykes N, Morris A, Weeks DE, Mott R, Lamb J, Bailey AJ, Monaco AP International Molecular Genetics Study of Autism Consortium (2005) Candidate-gene screening and association analysis at the autism-susceptibility locus on chromosome 16p: evidence of association at GRIN2A and ABAT. Am J Human Genet 76:950–966.

Belmonte MK, Bourgeron T (2006) Fragile X syndrome and autism at the intersection of genetic and neural networks. Nature Neurosci 9:1221–1225.

Blasi F, Bacchelli E, Pesaresi G, Carone S, Bailey AJ, Maestrini E (2006) Absence of coding mutations in the X-linked genes neuroligin 3 and neuroligin 4 in individuals with autism from the IMGSAC collection. Am J Med Genet B Neuropsychiatr Genet 141:220–221.

Boeckers TM, Bockmann J, Kreutz MR, Gundelfinger ED (2002) ProSAP/Shank proteins – a family of higher order organizing molecules of the postsynaptic density with an emerging role in human neurological disease. J Neurochem 81:903–910.

Chih B, Afridi SK, Clark L, Scheiffele P (2004) Disorder-associated mutations lead to functional inactivation of neuroligins. Human Mol Genet 13:1471–1477.

Chih B, Gollan L, Scheiffele P (2006) Alternative splicing controls selective trans-synaptic interactions of the neuroligin-neurexin complex. Neuron 51:171–178.

Chubykin AA, Atasoy D, Etherton MR, Brose N, Kavalali ET, Gibson JR, Südhof TC (2007) Activity-dependent validation of excitatory versus inhibitory synapses by neuroligin-1 versus neuroligin-2. Neuron 54:919–931.

Comoletti D, De Jaco A, Jennings LL, Flynn RE, Gaietta G, Tsigelny I, Ellisman MH, Taylor P (2004) The Arg451Cys-neuroligin-3 mutation associated with autism reveals a defect in protein processing. J Neurosci 24:4889–4893.

Comoletti D, Flynn RE, Boucard AA, Demeler B, Schirf V, Shi J, Jennings LL, Newlin HR, Südhof TC, Taylor P (2006) Gene selection, alternative splicing, and post-translational processing regulate neuroligin selectivity for beta-neurexins. Biochemistry 45:12816–12827.

Craig AM, Kang Y (2007) Neurexin-neuroligin signaling in synapse development. Curr Opin Neurobiol17:43–52.

Dean C, Scholl FG, Choih J, DeMaria S, Berger J, Isacoff E, Scheiffele P (2003) Neurexin mediates the assembly of presynaptic terminals. Nature Neurosci 6:708–716.

Dudanova I, Tabuchi K, Rohlmann A, Sudhof TC, Missler M (2007) Deletion of alpha-neurexins does not cause a major impairment of axonal pathfinding or synapse formation. J Comp Neurol 502:261–274.

Durand CM, Betancur C, Boeckers TM. Bockmann J, Chaste P, Fauchereau F, Nygren G, Rastam M, Gillberg IC, Anckarsäter H, Sponheim E, Goubran-Botros H, Delorme R, Chabane N, Mouren-Simeoni MC, de Mas P, Bieth E, Rogé B, Héron D, Burglen L, Gillberg C, Leboyer M, Bourgeron T (2007) Mutations in the gene encoding the synaptic scaffolding protein SHANK3 are associated with autism spectrum disorders. Nature genetics 39:25–27.

Fombonne E (2005) Epidemiology of autistic disorder and other pervasive developmental disorders. J Clin Psych 66 (Suppl 10):3–8.

Freitag CM (2007) The genetics of autistic disorders and its clinical relevance: a review of the literature. Mol Psych 12:2–22.

Frith U (1998) Cognitive deficits in developmental disorders. Scand J Psychol 39:191–195.

Gauthier J, Bonnel A, St-Onge J, Karemera L, Laurent S, Mottron L, Fombonne E, Joober R, Rouleau GA (2005) NLGN3/NLGN4 gene mutations are not responsible for autism in the Quebec population. Am J Med Genet B Neuropsychiatr Genet 132:74–75.

Graf ER, Zhang X, Jin SX, Linhoff MW, Craig AM (2004) Neurexins induce differentiation of GABA and glutamate postsynaptic specializations via neuroligins. Cell 119:1013–1026.

Graf ER, Kang Y, Hauner AM, Craig AM (2006) Structure function and splice site analysis of the synaptogenic activity of the neurexin-1 beta LNS domain. J Neurosci 26:4256–4265.

Guy J, Gan J, Selfridge J, Cobb S, Bird A (2007) Reversal of neurological defects in a mouse model of Rett syndrome. Science 315:1143–1147.

Hayashi ML, Rao BS, Seo JS, Choi HS, Dolan BM, Choi SY, Chattarji S, Tonegawa S (2007) Inhibition of p21-activated kinase rescues symptoms of fragile X syndrome in mice. Proc Natl Acad Sci USA 104:11489–11494.

Jacquemont ML, Sanlaville D, Redon R, Raoul O, Cormier-Daire V, Lyonnet S, Amiel J, Le Merrer M, Heron D, de Blois MC, Prieur M, Vekemans M, Carter NP, Munnich A, Colleaux L, Philippe A (2006) Array-based comparative genomic hybridisation identifies high frequency of cryptic chromosomal rearrangements in patients with syndromic autism spectrum disorders. J Med Genet 43:843–849.

Jamain S, Betancur C, Quach H, Philippe A, Fellous M, Giros B, Gillberg C, Leboyer M, Bourgeron T (2002) Linkage and association of the glutamate receptor 6 gene with autism. Mol Psych 7:302–310.

Jamain S, Quach H, Betancur C, Råstam M, Colineaux C, Gillberg IC, Soderstrom H, Giros B, Leboyer M, Gillberg C, Bourgeron T (2003) Mutations of the X-linked genes encoding neuroligins NLGN3 and NLGN4 are associated with autism. Nature Genet 34:27–29.

Kanner L (1943) Autistic disturbances of affective contact. Nerv Child 2:217–250.

Kim SA, Kim JH, Park M, Cho IH, Yoo HJ (2007) Family-based association study between GRIK2 polymorphisms and autism spectrum disorders in the Korean trios. Neurosci Res 58: 332–335.

Laumonnier F, Bonnet-Brilhault F, Gomot M, Blanc R, David A, Moizard MP, Raynaud M, Ronce N, Lemonnier E, Calvas P, Laudier B, Chelly J, Fryns JP, Ropers HH, Hamel BC, Andres C, Barthélémy C, Moraine C, Briault S (2004) X-linked mental retardation and autism are associated with a mutation in the NLGN4 gene, a member of the neuroligin family. Am J Human Genet 74:552–557.

Manning MA, Cassidy SB, Clericuzio C, Cherry AM, Schwartz S, Hudgins L, Enns GM, Hoyme HE (2004) Terminal 22q deletion syndrome: a newly recognized cause of speech and language disability in the autism spectrum. Pediatrics 114:451–457.

Meyer G, Varoqueaux F, Neeb A, Oschlies M, Brose N (2004) The complexity of PDZ domain-mediated interactions at glutamatergic synapses: a case study on neuroligin. Neuropharmacology 47:724–733.

Miles JH, Hillman RE (2000) Value of a clinical morphology examination in autism. Am J Med Genet 91:245–253.

Missler M (2003) Synaptic cell adhesion goes functional. Trends in neurosciences 26:176–178.

Motazacker M, Rost BR, Hucho T et al. (2007) A defect in the ionotropic glutamate receptor 6 gene (GRIK2) is associated with autosomal recessive mental retardation. Am J Human Genet, in press.

Prange O, Wong TP, Gerrow K, Wang YT, El-Husseini A (2004) A balance between excitatory and inhibitory synapses is controlled by PSD-95 and neuroligin. Proc Natl Acad Sci USA 101:13915–13920.

Premak J, Woodruff G (1978) Does the chimpanzee have a theory of mind? Behav Brain Sci 4:515–526.

Scheiffele P, Fan J, Choih J, Fetter R, Serafini T (2000) Neuroligin expressed in nonneuronal cells triggers presynaptic development in contacting axons. Cell 101:657–669.

Sebat J, Lakshmi B, Malhotra D, Troge J, Lese-Martin C, Walsh T, Yamrom B, Yoon S, Krasnitz A, Kendall J, Leotta A, Pai D, Zhang R, Lee YH, Hicks J, Spence SJ, Lee AT, Puura K, Lehtimäki T, Ledbetter D, Gregersen PK, Bregman J, Sutcliffe JS, Jobanputra V, Chung W, Warburton D, King MC, Skuse D, Geschwind DH, Gilliam TC, Ye K, Wigler M (2007) Strong association of de novo copy number mutations with autism. Science 316:445–449.

Shaywitz SE, Shaywitz BA, Fletcher JM, Escobar MD (1990) Prevalence of reading disability in boys and girls. Results of the Connecticut Longitudinal Study. JAMA 264:998–1002.

Shuang M, Liu J, Jia MX, Yang JZ, Wu SP, Gong XH, Ling YS, Ruan Y, Yang XL, Zhang D (2004) Family-based association study between autism and glutamate receptor 6 gene in Chinese Han trios. Am J Med Genet B Neuropsychiatr Genet 131:48–50.

Strutz-Seebohm N, Korniychuk G, Schwarz R, Baltaev R, Ureche ON, Mack AF, Ma ZL, Hollmann M, Lang F, Seebohm G (2006) Functional significance of the kainate receptor GluR6(M836I) mutation that is linked to autism. Cell Physiol Biochem 18:287–294.

Szatmari P, Paterson AD, Zwaigenbaum L, Roberts W, Brian J, Liu XQ, Vincent JB, Skaug JL, Thompson AP, Senman L, Feuk L, Qian C, Bryson SE, Jones MB, Marshall CR, Scherer SW, Vieland VJ, Bartlett C, Mangin LV, Goedken R, Segre A, Pericak-Vance MA, Cuccaro ML, Gilbert JR, Wright HH, Abramson RK, Betancur C, Bourgeron T, Gillberg C, Leboyer M, Buxbaum JD, Davis KL, Hollander E, Silverman JM, Hallmayer J, Lotspeich L, Sutcliffe JS, Haines JL, Folstein SE, Piven J, Wassink TH, Sheffield V, Geschwind DH, Bucan M, Brown WT, Cantor RM, Constantino JN, Gilliam TC, Herbert M, Lajonchere C, Ledbetter DH, Lese-Martin C, Miller J, Nelson S, Samango-Sprouse CA, Spence S, State M, Tanzi RE, Coon H, Dawson G, Devlin B, Estes A, Flodman P, Klei L, McMahon WM, Minshew N, Munson J, Korvatska E, Rodier PM, Schellenberg GD, Smith M, Spence MA, Stodgell C, Tepper PG, Wijsman EM, Yu CE, Rogé B, Mantoulan C, Wittemeyer K, Poustka A, Felder B, Klauck SM, Schuster C, Poustka F, Bölte S, Feineis-Matthews S, Herbrecht E, Schmötzer G, Tsiantis J, Papanikolaou K, Maestrini E, Bacchelli E, Blasi F, Carone S, Toma C, Van Engeland H, de Jonge M, Kemner C, Koop F, Langemeijer M, Hijimans C, Staal WG, Baird G, Bolton PF, Rutter ML, Weisblatt E, Green J, Aldred C, Wilkinson JA, Pickles A, Le Couteur A, Berney T, McConachie H, Bailey AJ, Francis K, Honeyman G, Hutchinson A, Parr JR, Wallace S, Monaco AP, Barnby G, Kobayashi K, Lamb JA, Sousa I, Sykes N, Cook EH, Guter SJ, Leventhal BL, Salt J, Lord C, Corsello C, Hus V, Weeks DE, Volkmar F, Tauber M, Fombonne E, Shih A (2007)

Mapping autism risk loci using genetic linkage and chromosomal rearrangements. Nature Genet 39:319–328.
Talebizadeh Z, Lam DY, Theodoro MF, Bittel DC, Lushington GH, Butler MG (2006) Novel splice isoforms for NLGN3 and NLGN4 with possible implications in autism. J Med Genet 43:e21.
Tomblin JB, Records NL, Buckwalter P, Zhang X, Smith E, O'Brien M (1997) Prevalence of specific language impairment in kindergarten children. J Speech Lang Hear Res 40:1245–1260.
Tuchman R, Rapin I (2002) Epilepsy in autism. Lancet Neurol 1:352–358.
Varoqueaux F, Jamain S, Brose N (2004) Neuroligin 2 is exclusively localized to inhibitory synapses. Eur J Cell Biol 83:449–456.
Varoqueaux F, Aramuni G, Rawson RL, Mohrmann R, Missler M, Gottmann K, Zhang W, Südhof TC, Brose N (2006) Neuroligins determine synapse maturation and function. Neuron 51:741–754.
Vincent JB, Kolozsvari D, Roberts WS, Bolton PF, Gurling HM, Scherer SW (2004) Mutation screening of X-chromosomal neuroligin genes: no mutations in 196 autism probands. Am J Med Genet B Neuropsychiatr Genet 129B:82–84.
Vorstman JA, Staal WG, van Daalen E, van Engeland H, Hochstenbach PF, Franke L (2006) Identification of novel autism candidate regions through analysis of reported cytogenetic abnormalities associated with autism. Mol Psych 11:1, 18–28.
Yan J, Oliveira G, Coutinho A, Yang C, Feng J, Katz C, Sram J, Bockholt A, Jones IR, Craddock N, Cook EH Jr, Vicente A, Sommer SS (2004) Analysis of the neuroligin 3 and 4 genes in autism and other neuropsychiatric patients. Mol Psych 10: 329–32.
Ylisaukko-oja T, Rehnstrom K, Auranen M, Vanhala R, Alen R, Kempas E, Ellonen P, Turunen JA, Makkonen I, Riikonen R, Nieminen-von Wendt T, von Wendt L, Peltonen L, Järvelä I (2005) Analysis of four neuroligin genes as candidates for autism. Eur J Human Genet 13:1285–1292.

Subject Index

List of previously published volumes in the series "Research and Perspectives in Alzheimer's Disease"

A. Pouplard-Barthelaix et al. (Eds.) (1988) Immunology and Alzheimer's Disease

P.M. Sinet et al. (Eds.) (1988) Genetics and Alzheimer's Disease

F. Gage et al. (Eds.) (1989) Neuronal Grafting and Alzheimer's Disease

F. Boller et al. (Eds.) (1989) Biological Markers of Alzheimer's Disease

S.I. Rapoport et al. (Eds.) (1990) Imaging, Cerebral Topography and Alzheimer's Disease

F. Hefti et al. (Eds.) (1991) Growth Factors and Alzheimer's Disease

Y. Christen and P.S. Churchland (Eds.) (1992) Neurophilosophy and Alzheimer's Disease

F. Boller et al. (Eds.) (1992) Heterogeneity of Alzheimer's Disease

C.L. Masters et al. (Eds.) (1994) Amyloid Protein Precursor in Development, Aging and Alzheimer's Disease

K.S. Kosik et al. (Eds.) (1995) Alzheimer's Disease: Lessons from Cell Biology

A.D. Roses et al. (Eds.) (1996) Apolipoprotein E and Alzheimer's Disease

B.T. Hyman et al. (Eds.) (1997) Connections, Cognition and Alzheimer's Disease

S.G. Younkin et al. (Eds.) (1998) Presenilins and Alzheimer's Disease

R. Mayeux and Y. Christen (Eds.) (1999) Epidemiology of Alzheimer's Disease : From Gene to Prevention

V.M.-Y. Lee et al. (Eds.) (2000) Fatal Attractions: Protein Aggregates in Neurodegenerative Disorders

K. Beyreuther et al. (Eds.) (2001) Neurodegenerative Disorders: Loss of Function Through Gain of Function

A. Israël et al. (Eds.) (2002) Notch from Neurodevelopment to Neurodegeneration: Keeping the Fate

D.J. Selkoe, Y. Christen (Eds.) (2003) Immunization Against Alzheimer's Disease and Other Neurodegenerative Disorders

B. Hyman et al. (Eds.) (2004) The Living Brain and Alzheimer's Disease

J. Cummings et al. (Eds.) (2005) Genotype – Proteotype – Phenotype relationships in Neurodegenerative Disease

M. Jucker et al. (Eds.) (2006) Alzheimer: 100 Years and Beyond